COMMUNICATION FOR KINESIOLOGY

COMMUNICATION FOR KINESIOLOGY

A Practical Guide

Jason Holt

CANADIAN SCHOLARS
Toronto | Vancouver

Communication for Kinesiology: A Practical Guide
Jason Holt

First published in 2022 by
Canadian Scholars, an imprint of CSP Books Inc.
425 Adelaide Street West, Suite 200
Toronto, Ontario
M5V 3C1

www.canadianscholars.ca

Library and Archives Canada Cataloguing in Publication

Title: Communication for kinesiology : a practical guide / Jason Holt.
Names: Holt, Jason, 1971- author.
Description: Includes bibliographical references and index.
Identifiers: Canadiana (print) 20220177139 | Canadiana (ebook) 20220177171 |
ISBN 9781773383101 (softcover) | ISBN 9781773383125 (PDF) |
ISBN 9781773383118 (EPUB)
Subjects: LCSH: Kinesiology—Authorship. | LCSH: Sports medicine—Authorship. |
LCSH: Communication.
Classification: LCC R119 .H65 2022 | DDC 808.06/661—dc23

Cover design by Rafael Chimicatti
Cover image by Shutterstock/prarinyafoto
Page layout by S4Carlisle Publishing Services

21 22 23 24 25 5 4 3 2 1

Printed and bound in Ontario, Canada

Canada

Contents

Acknowledgements

Thanks to everyone at Canadian Scholars, especially Sarah Powell, Katherine Kurowski, and Lindsey Simeonidis. Their enthusiasm and guidance were invaluable. Several anonymous reviewers gave detailed and helpful feedback for revising the manuscript. My colleague Gabriela Tymowski-Gionet provided a wealth of encouraging feedback. Thanks to Cindy Angelini for the careful copyediting. Thanks last to all the kinesiology students who have or had to take my first-year communication course. No less than it is intended for you, this book was inspired by you.

Note to Instructors

This book is deliberately modular, with each chapter standing more or less on its own without presupposing material presented earlier. This gives instructors the freedom to select and order topics covered in whatever way best fits their specific course and personal teaching style. For some instructors, for instance, it might make sense to cover giving presentations (chapters 11 and 12) earlier in the course or to defer discussion of basic theory (chapter 1) until later. As I see it, flexibility is as much a virtue here as it is in athletics.

Preface

Welcome to this book. In putting it together, I have drawn on over 15 years' experience teaching communication to kinesiology students and over 20 years teaching at the university level. Think of yourself as in training and me as your coach. Just like a good coach in athletics, I want to help you to take your game to the next level. I want to help you succeed both as a student and, later on, as a professional. For this to work, heightened attention to detail and willingness to change your habits are essential. You need to put in the effort.

This book assumes a contextual approach to communication, focusing on formal writing and presentations in scholarly and professional settings. The advice contained in these pages is, for the most part, practical but theoretically grounded as well, sometimes explicitly. Changing how you think about the communication process can motivate you to hone your writing and presentation skills. The idea is *not* that your communication practices need a complete overhaul. This book is not about changing your everyday writing and speech, but about learning to play a certain sort of game better, as with a favourite sport or leisure activity. Taking a degree or being a professional in kinesiology is, in a sense, playing a game, which means adapting to the rules and conventions in effect for the sake of playing the game *well*. That is why I encourage you to see yourself as in training and me as your coach. If you put in the work, I can help you take your communication game to the next level.

CHAPTER 1

Basic Theory

CHAPTER OVERVIEW

This chapter introduces some basic communication theory. We begin by exploring two different ways of thinking about what communication is, followed by the five elements of communication. Next we explore how communication varies given the number of people involved and the tricky verbal/nonverbal distinction. We end by briefly considering the hard question of how we ever really know whether we have communicated successfully at all.

PARADIGMS

In the sci-fi movie *Contact*, starring Jodie Foster and based on the novel by Carl Sagan, a team involved in SETI (the Search for Extraterrestrial Intelligence), using a powerful radio telescope, receives a possible signal from space. Their interest is first piqued by the phenomenon's strong, regular pulse standing out from the weak, chaotic background noise. Is it a message from aliens? Not necessarily. Unintelligent natural phenomena, from heartbeats to certain stars (pulsars), can produce pulses, as can machines, although machines imply intelligence in design. At one point the pulse ceases, then resumes, and the team notices that the pulses are coming in groups: at first two, then three, then five, seven, eleven ... the series of prime numbers—those numbers evenly divisible only by themselves and one. This finally convinces them that what they are dealing with, what they are receiving, is a genuine signal from an alien intelligence.

This scene from *Contact* will serve as a touchstone for our discussion in this chapter of the two main paradigms of communication. Before this, however, it will help to say something about what paradigms are. The term **paradigm** was popularized by the philosopher of science Thomas Kuhn (1970), and it means one of two things. *Paradigm* is sometimes used to refer to model examples, as Lamborghini is a paradigm sports car and LeBron James is a paradigm basketball player. The term is also used to refer to a theoretical framework, a way of seeing the world. In this sense, we also speak of paradigm *shifts*, major changes in theoretical perspective, such as, in physics, the move from Newtonian classical mechanics to Einsteinian relativity. Both meanings of this term, though the latter more so, pertain to the paradigms of communication.

The Transmission Paradigm

When people think of communication, especially the verb *communicate*, they usually have the **transmission paradigm** in mind. To communicate in this sense is to *convey*. The verb often has a direct object (the information that is being conveyed) and an indirect object (the person meant to receive it). If you understand the sentence that you are currently reading, this means that, by the time you finish it, I will have successfully communicated a thought *to* you. For a symbol of this paradigm, think of a radio tower (figure 1.1). A radio transmits information encoded in radio waves to the antenna of your car stereo, where that signal is decoded into the original sound. In the same way, I have encoded my thoughts into these sentences, which are transmitted visually and then decoded by your brain. The communication is successful when you have decoded the message I encoded in those sentences, correctly interpreting the meaning I intended to express.

This would seem to be communication in its most basic, primordial form, a concept that has been around for millennia at least. According to Radford (2005), however, we can trace this paradigm to a surprisingly recent origin, the 17th-century philosopher John Locke. Radford claims "it was Locke's use of the

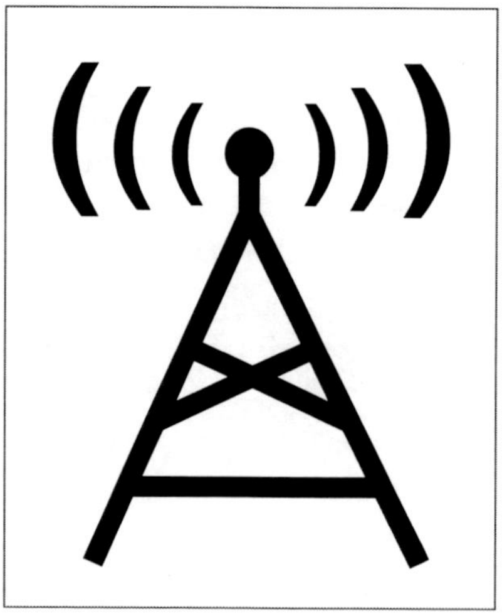

Figure 1.1. The Transmission Paradigm
Source: Jason Holt

term 'communication' in his *Essay* [*Concerning Human Understanding*] that was to form the decisive shift to the modern sense of the term" (p. 18). In Locke's view, "to make words serviceable to the end of communication, it is necessary ... that they excite in the hearer exactly the same idea they stand for in the mind of the speaker. Without this, [people] ... convey not thereby their thoughts" (1836, p. 350). According to the transmission paradigm, therefore, thoughts that are essentially private are encoded in publicly accessible speech and other signals, only to be decoded into the private realm of an interpreter's thoughts.

To anticipate points of contrast between the two paradigms, two further remarks about the transmission paradigm are in order. First, the transmission paradigm is *mechanistic*, as it presents the radio tower's transmission to a car stereo antenna and you talking with a friend over coffee as essentially the same type of communicative process: information going from source to recipient. Second, it is *unidirectional*, analyzing communication as sender-to-receiver and not vice versa. Of course, when you and a friend talk over coffee, there will be a lot of back-and-forth interaction; else it would hardly be a conversation. A message that responds to another is called **feedback**. You and your friend will often respond to what the other person has said. But according to the transmission model, such feedback is itself a sender-to-receiver process in its own right. In a two-person conversation, the roles have simply switched and will often switch back and forth many times.

The Sharing Paradigm

The second paradigm construes communication rather differently. It is sometimes referred to as a ritual view of communication (e.g., Carey, 1992), although we will call it the **sharing paradigm**. This is not a mere terminological preference, as *ritual* can be a misleading term, suggesting that communication must be, like rituals, a matter of repetition or habit, or must happen in a certain institutional setting. Neither impression is true, and both obscure what for the sharing paradigm is the essence of communication. To communicate in this sense is not to convey but to *commune*. The emphasis likewise is not on the messages that pass between senders and receivers, but on the people themselves. As an icon of this paradigm, think of people holding hands (figure 1.2). This is communication *with*, a sharing of meaningful experience. Here the emphasis shifts from information exchange to personal and social meaning.

Because of this shift of emphasis, the sharing paradigm distinguishes itself in other ways as well. First, because it is mechanistic, the transmission paradigm takes as its central image and reduces human interaction to the literal information transmission of machines. The sharing paradigm, by contrast, is not mechanistic but *humanistic*. We are interested in communication in terms of meaningful human interactions, and this implies a different level of analysis: psychological and social,

Figure 1.2. The Sharing Paradigm
Source: Adapted from Flaticon.com

not mechanistic or information-theoretic. Second, whereas the transmission paradigm implies the asymmetry of a sender-to-receiver process, the sharing paradigm instead emphasizes commonality, not difference. In terms of the transmission paradigm, a university lecture is a message imparted by a source (professor) to receivers (students). Yet in terms of the sharing paradigm, what matters is not who is lecturing to whom, but that, whatever the respective roles, the people involved (professor and students) are sharing in a meaningful experience (the lecture).

Paradigms of Communication

- Transmission
    - Informational object
    - Unidirectional focus
    - Mechanistic analysis
- Sharing
    - Meaningful object
    - Nondirectional focus
    - Humanistic analysis

Which Paradigm?

The transmission and sharing paradigms represent different perspectives on communication. We might ask ourselves which of the two, if either, is correct. If we take the paradigms as presenting different definitions of communication, of what does and does not count as communication, then we have a genuine controversy. Each approach includes some form of communication excluded by the other approach. Consider someone sharing a comfortable silence with their husband or wife.

Because there is no exchange of information, the comfortable silence is ruled out by the transmission view. Even so, however, this is precisely the sort of shared meaningful experience that the sharing paradigm includes as a significant form of communication. Likewise, mere data transfer, say, between fully automated computers, seems to count as communication within the transmission paradigm only. The moral? To commune is not to convey, and to convey is not to commune.

As another illustration of the difference between these paradigms, think of how a sports medicine doctor or athletic trainer may interact with an injured athlete. The specialist might be able to diagnose an injury correctly and prescribe the appropriate treatment, and so transmit the relevant information to the athlete. But this does not mean that the specialist has a good bedside manner, which implies a certain level of respectful interaction. An incompetent specialist could interact well with clients on a personal level though fail to serve their professional role. Ideally, we want professionals who communicate well in terms of what *each* paradigm emphasizes, that is, both on an informational and on a personal level.

Where does this leave us? Suppose we agree about what is included and excluded from the communication category. We could reinterpret the comfortable silence example as an interesting type of signal in its own right (as in "No news is good news"), and we might redescribe automated data transfer in terms of the meaning-sharing intentions of the programmers. Despite this possibility, each paradigm still seems to lay exclusive claim to capturing communication at its very core, its essence. Should the flag of this disputed territory be mechanistic? Humanistic? Or is this impression also misleading? Perhaps it is analogous to the question of solidity in physics. My desk is solid. From a normal (i.e., macrophysical) perspective, it contains no empty space. But from a microphysical perspective, it is *mostly* empty space. We can dispel the contradiction by noting that we are talking about different levels of analysis.

By this analogy, then, each paradigm describes a different level of analysis for studying communication rather than describing a different type of communication. Most of the communication phenomena we consider will be usefully describable either in terms of information transfer or meaning sharing. Things at lower levels of analysis are said to implement or realize those at higher levels. In this sense, we often say that the mind is realized by the brain, that psychological processes are implemented by, or occur via, brain processes. In a similar way, we can combine both paradigms into a single unified definition: **communication** is the sharing of meaning via information transmission. We need not exclusively choose one paradigm over the other when, seen correctly, this proves a false dichotomy.

Definition of Communication

Communication is the sharing of meaning via information transmission.

The balance between paradigms implied by the above definition has not just theoretical but also practical dimensions. Communication can go awry when one aspect is overemphasized to the neglect of the other. A doctor with a poor bedside manner neglects the sharing paradigm, and an engaging but unprepared lecturer neglects the transmission paradigm. When we ask someone, "How are you?" we usually want the social exchange, not the information, though sometimes we do want the information, in which case the socially expected "Fine" will become an unsatisfying answer. Any communicative situation may be seen as having different corresponding degrees of appropriate emphasis on its informational and personal or social aspects.

FIVE ELEMENTS

We shift gears now from discussing different theoretical lenses through which to view communication to analyzing its constituents. Communication comprises different elements that are important to distinguish. These elements are necessary in that whenever communication is happening, each element is present and plays an identifiable role in shaping what is going on. For theoretical as well as practical reasons, we will distinguish five such necessary elements (see, e.g., Dobkin & Pace, 2006).

First, there are the **agents**: the subjects involved, those who send or receive information, those who attempt to share meaning or meaningful experience. Since we are interested primarily in human communication, we will normally assume that the agents involved are people. This is just a convenience, however. Animals communicate with one another, sometimes with humans, and the aliens in *Contact*, if real, would be nonhuman communicators.

Next, there are the **messages**: the units of information that pass between, or the meaningful items shared among, the agents involved. These messages can range from small nonverbal cues, like a nod of greeting, to long complex messages, like a book. We can often discern what is and what is not part of a particular message (email, speech, what have you), though not necessarily always (e.g., in face-to-face interactions). If you receive a smudged handwritten note on lined yellow paper, you know that the smudges, the lines, and the yellow of the paper are not part of the text itself. What seem to be smudges on a child's drawing, however, might or might not be elements of the drawing itself. Likewise, someone's facial expression in conversation could, but need not, have anything to do with what they are trying to say.

Our third element, **channels**, is a matter of how a message gets through, how information is transmitted, how meaning is shared. Think less in terms of "the media" (television or YouTube channels) and more, and more broadly, in terms of the medium or media in which a message is expressed. When I give a lecture to a class, often with PowerPoint slides, various overlapping channels are identifiable: auditory, visual, textual, nonverbal. Channels in this sense are a matter of what senses are used and what type of stimulus is coming through. A photograph is visual but not linguistic, whereas text is both visual and linguistic. The visual channel includes these among other subtypes.

Another element is **noise**. For a paradigm case here, think of telephone static. Sometimes noise is understood as anything that interferes with a message getting through (see, e.g., Dobkin & Pace, 2006). Although telephone static, depending on the amount, certainly does interfere with a message getting through, such a definition is too broad. If death cuts off a conversation mid-sentence, that interferes with the message getting through, but it is not noise. Noise is *channel-specific* message interference. Channel-specificity is tied to sensory input. If you listen to someone on the phone, someone playing loud music in the next room may count as noise just as static on the line would.

The fifth element is **context**: the environment in which information is being transmitted and meaning shared, which has relevant physical, social, and cultural dimensions. It is usually far more acceptable, for instance, to use rougher language in certain social settings with friends than in more formal or professional settings. In some cultures, eye contact is seen as a sign of respect, whereas in other cultures it is seen as a sign of *dis*respect.

Elements of Communication

- *Agents*: people involved
- *Messages*: what is communicated
- *Channels*: how messages get through
- *Noise*: channel-specific message interference
- *Context*: communicative setting

BY THE NUMBERS

Among various ways of distinguishing different types of communication, one of the most useful depends on the number of agents involved. Vary the number of people, and relevant phenomena are liable to change in crucial—sometimes obvious, sometimes subtle—respects.

What is the minimum number of people needed for communication to occur? It might seem that you need at least two to tango: one sender or sharer, one receiver or "sharee." This is true in that both roles need to be filled, but that does not necessitate two different people, since both roles may be played by a single person. The minimum number, then, is one. Communicating with oneself is called **intrapersonal communication**. Everyday examples include writing your own grocery list, talking to yourself silently or aloud, or taking class notes.[1] Adding another person yields **interpersonal communication**, which is communication between exactly two agents. Here emerge differences in perspective that increase the likelihood of misunderstanding but also that of acquiring knowledge and developing relationships.

The boundaries may seem arbitrary, but the next communicative type, **small group communication**, pertains to groups of three to seven people (Dobkin & Pace, 2006). The dynamics change when you add a third to the mix. Think of the difference between going for coffee with one friend and going for coffee with two. Behaviours change, as do styles and means of interaction. This might be intuitive enough, but why should the upper limit be seven? The answer is that between seven and eight agents, the dynamics change again. Someone with a fear of public speaking will often be significantly more comfortable interacting in a group of seven or fewer. There are noticeable differences in heart rate, for example (Porhölä et al., 1993). The fact that the threshold lies between seven and eight is far clearer than why this should be so.

When the number of agents involved increases to eight or more, we enter the realm of **public communication**. Think of public speaking, newsletters, social media, music concerts, institutional memoranda, and so on, where the number of people involved are substantial, though not necessarily vast. Where the numbers involved are particularly large, public communication becomes **mass communication**: in effect, public communication gone viral, with great numbers of people and technology to mediate message delivery. Think here of viral videos and internet-available news items, as well as more traditional broadcast media (television, radio) and even print publishing (books, magazines).

Types of Communication (Distinguished by Number of People)

- Intrapersonal (one person)
- Interpersonal (two people)
- Small group (three to seven people)
- Public (more than seven people)
- Mass (large numbers of people)

THE VERBAL/NONVERBAL DISTINCTION

The word *verbal* is ambiguous. **Verbal communication** can mean either spoken language or language generally. In the first sense, writing counts as nonverbal, but in the second it counts as verbal. Similarly, sign language is nonverbal in that it is unspoken, but verbal in that it is a form of language, like English, with a systematic vocabulary and grammar. What we often call "body language" is not strictly language in this sense at all. But as a type of nonverbal communication, it is meaningful, and such **nonverbal cues** are important in face-to-face communication. In some sense, they even intrude into written communication. For example, the use of italics in writing for emphasis can help convey *tone*, which modulates vocalized language but is not strictly verbal in either sense.

Ambiguity of *Verbal* and *Nonverbal*

- *Verbal* can mean
  - Spoken
  - Linguistic
- *Nonverbal* can mean
  - Unspoken
  - Nonlinguistic

Nonverbal cues have certain properties that should be acknowledged. By definition, they are nonlinguistic. They are also given off *continuously* throughout a person's waking life (Dobkin & Pace, 2006). Nonverbal cues are typically, if not always, tied to emotional states, given off unintentionally, and highly contextual (Dobkin & Pace, 2006). A simple nonverbal cue such as maintaining eye contact can vary in meaning depending on the situation, the relationship involved, the cultural context, and the gender of those involved. Eye contact could convey interest in one situation and hostility in another, be perceived as attentive between friends but creepy between strangers, show a sign of respect in one culture but *dis*respect in another, and communicate something different from a woman than from a man.

Nonverbal cues can be classified broadly under two subtypes depending on how they are accessed by interpreters. Visual nonverbal cues include eye contact, facial expressions, and gestures, as well as **proxemics** and **kinesics** (Dobkin & Pace, 2006). Proxemics—think proximity—is the meaning of distance or changes in distance: leaning forward when paying attention, for instance, or the intimacy of being in someone's personal space. Kinesics, by contrast, is the nonverbal use of movement to communicate, as in dance. Distinct from these, we may identify various types of nonvisual cues. Other sense modalities include hearing (**vocalics**), touch (**haptics**), and smell ("**olfactics**") (Dobkin & Pace, 2006).[2] Other nonverbal signifiers include time (**chronemics**) and made objects (**artifactics**; Dobkin & Pace, 2006). Being punctual is a chronemic cue, as is spending time with someone. Think of how we interpret artifacts such as wedding rings, nice or shabby clothes, or a sports car driven by a middle-aged man. On the sense modality side, a pat on the back and a handshake are haptic cues, and the wearing of perfume or cologne is an olfactic cue. As for vocalics, they might be pure (e.g., crying out in pain) or **paralinguistic cues**, that is, not linguistic in themselves but modulating spoken language (e.g., vocal tone, cadence, and volume), a matter of how something is said rather than what is being said. There are in fact various interesting ways in which nonverbal cues may interact with verbal ones, sometimes in concert, sometimes in conflict.[3] Think of people using hand gestures when they speak, or people whose facial expressions contradict how they state they are feeling.

Types of Nonverbal Cues

- Visual
  - Eye contact
  - Facial expressions
  - Gestures
  - Proxemics (distance-based)
  - Kinesics (movement-based)
- Nonvisual
  - Vocalics (voice-based)
    - Nonlinguistic
    - Paralinguistic (modifying speech)
  - Haptics (touch-based)
  - Olfactics (smell-based)
  - Chronemics (time-based)
  - Artifactics (object-based)

HOW DO YOU KNOW?

In his memoir *On Writing* (2000), Stephen King, in a chapter titled "What Writing Is," makes the provocative claim that writing is "telepathy, of course" (p. 103). This tongue-in-cheek assertion succinctly encapsulates both the intuitive appeal and the susceptibility to skepticism of the transmission paradigm.

So far, we have considered theoretical approaches that take it for granted that communication happens at all. But suppose we adopt a skeptical attitude, casting doubt on whether communication ever really occurs or, if it does, asking how we verify that it does. How do we know that we have communicated successfully? How can anyone tell, for instance, that the thoughts you glean from reading this chapter are equivalent to the thoughts that prompted me to write it? We cannot simply compare our respective thoughts the way we might compare paint or fabric swatches, since each of us experiences only one particular mind: our own.[4] This doubt may seem to apply solely to the transmission model, since an idea that fails to survive transmission is not transmitted at all. But such doubts also apply to the sharing model. Sharing a meaningful experience does not necessarily involve experiencing it the same way or deriving the same meaning from it.

What are we to make of this? Part of the answer is that *mis*communication often reveals itself. For instance, if you and a friend agree to watch "that inspiring sport film," you might start streaming *Rocky* only to have your friend say, "Oh, I thought you meant *Rudy*." The more feedback between or among people, the greater the interaction, and the more likely such misunderstanding will be revealed. But it is not as though successful communication is simply the lack of miscommunication. More important is that a lack of shared concepts and experiences matters less than we may think it does. Our cat, Pushkin, is perfectly able to signal when he wants to go outdoors, though his concept of the outdoors is radically different from mine. When I let Pushkin out, this is not mere behavioural compliance. If I fail to open the door, his facial expression is the feline equivalent of the famous line from *Cool Hand Luke*: "What we've got here is failure to communicate." I understand the message, just as *Contact*'s SETI team knows that the mathematical pulse signals an alien intelligence. It is in the coordination of thought and action that we find communication.

BRIEF RECAP

In this chapter, we introduced some basic communication theory:

- Paradigms of communication
- Five elements of communication
- Types of communication (distinguished by number of people)
- The verbal/nonverbal distinction
- Communication skepticism

CHAPTER 2

Composing Messages

CHAPTER OVERVIEW

In this chapter, we will explore the creation of messages in written or spoken form and in formal or informal contexts. We begin by examining the process of composition and useful ways to think about it, including preparation and motivation. Then we look at how to interpret different kinds of advice about writing: which principles should be considered strict(ish) rules and which more as rough guidelines. Finally, we discuss the significance of structure in compositions, whether that structure is explicit, as in a lab report, or implicit, as in a formal email.

WHAT COMPOSITION IS

It may seem pretentious to think of talking with someone or writing something down as **composition**. *Composition* suggests that you are doing something akin to a great creator of music, such as Mozart. A composer, in that sense, creates amazing and beautiful works using something besides language as a medium. We might consider great writers like Shakespeare to compose works, and we may even understand how great coaches or teachers are composers of the speeches they deliver to their team or class. But we think of ordinary speech and writing differently.

We may have trouble thinking of writing an essay, much less an email to a friend or professor, as composition. But the creation of messages, however simple, however ordinary, counts as composition, and thinking of it that way may help us to write and speak more effectively. We tend to work harder on things we care about, and if we see our active contribution to communication as the composition of messages, we just might be motivated to care a little more about what we are doing, and thus do a better job.

THE PROCESS (AND HOW TO THINK ABOUT IT)

The writing process can be daunting, especially with unfamiliar or time-consuming assignments. However, the more you know about the writing process and how to frame it in your mind, the easier it will become. In this section, we will cover the writing process, from setting the stage to polishing the final draft. You have already done a lot of writing in your life, perhaps far too much for your taste. Some of the advice I give may be familiar, or even obvious, to you. But there should be a few new tips to help you along.

Setting the Stage

It is natural to think that the key to getting writing done is just to do it. In a sense, this is true, but in another sense, it is misleading. Setting the stage for writing is almost as important to the writing as setting the stage for a stage play is to performing the play. It would be a stretch to say that the writing begins before you begin to write, but the writing context will affect the writing process even before fingers touch keyboard or pen touches paper. You may have a lot or very little control of the writing context. However, exercising whatever control you have is important. Be self-aware and experiment.

1. *Give yourself time.* Even with the mundane task of writing a grocery list, you may find it useful to take a few minutes, sit down with pen and paper, and think about what you want to put on the list and in what order. The time you set aside for a writing task will obviously depend on the type of task and on the task's appropriate or required length. For longer assignments, try giving yourself not just an appropriate amount of time but *extra* time, as well as starting early. To quote the *Tao Te Ching*, "Begin your task before it becomes a burden" (Lao Tzu, 2008, p. 83).[1]
2. *Attend to your writing space.* Some people write well at a dedicated workspace at home, such as a desk, because it is private and quiet. Others prefer quiet but public spaces like libraries. Still others write better in busier public spaces like coffee shops. No matter the location, some people write better with a tidy workspace, others with a messy space. These preferences can differ not only by personal temperament but also by type of writing task and even time of day.
3. *Have your materials at hand.*[2] This includes more than the primary writing materials you use, whether you write on a laptop or with pen and paper. It also includes whatever resource materials you need, such as instructions, notes, books, or PDFs of articles. Along with these primary and secondary materials, you should gather ancillary items that might help you think and write, such as a cup of coffee, a lucky hat, a snack, or background music.

4. *Develop a routine.* Drawing on points 1 to 3 above, consider deliberately crafting a writing routine. This will involve writing regularly, at a designated time, in a special place, with comfortingly familiar materials at hand. This strategy will help you get a lot of writing done not only when you have many different tasks to juggle, but also and especially when you face major assignments like term papers and honours or graduate theses. If an established routine turns unproductive, though, be experimental and change things up.

Getting Started, Staying Motivated

Getting started can be the hardest part of any writing assignment. It helps to make sure you know exactly what is expected of you, what sort of text you are being asked to produce, and something of the nature of that kind of text. You will not make much headway on an annotated bibliography if you are unaware what it means to annotate a bibliography. Assuming you know what the task is and, more narrowly, the further instructions to limit and guide your work, you may still find it difficult to get started. Motivation can be a problem, sometimes because you do not care enough about the assignment, and sometimes because you care too much. Ideally, you want to find yourself in the middle of these extremes: caring enough about the assignment to get to work on it but not caring so much that you become paralyzed by perfectionism—which can be a genuine problem among conscientious students. Two ways of thinking about and addressing the problem of motivation are getting to it and sticking with it.

One approach you are probably familiar with already is to set up a reward for yourself for getting a certain amount of work done. Maybe you want to check your email or go for a workout, and you decide that you will only do so once you have produced a paragraph or a page. Deferred gratification is a useful exercise in self-control. In the middle of your work, you could want, say, another cup of coffee. To give yourself an extra push before you allow yourself the caffeine kick, give yourself another goal to reach first, so the coffee carrot will be dangling in your mind as a reward for meeting the objective. A variation on this method is what we might call **creative procrastination**. If you have two unappealing tasks, A and B, and the thought of doing A is more off-putting than the thought of doing B, then try doing B *as a way of avoiding doing A.* You may be surprised how quickly you rush to work on a writing assignment when you think that there is something even *less* pleasant awaiting you.

Another approach is to find value in the work itself, rather than as a consequence of getting work done. You might challenge yourself, for instance, to write a tight paragraph in the next half-hour or to write a page before you get up from your desk. Many kinesiology students have experience taking on challenges, and if they can mentally frame all or part of a writing task as a challenge, that can help motivate them. By doing this, you are turning the writing process from work into play. Indeed, you can even turn the writing task into a series of little games, or one extended game. One quirky game that works for me is to try to produce a text that looks close to fully justified even though it is only left-justified with a "ragged" edge. This makes the writing process a little longer, but it has various advantages.[3] More importantly, because I turn the writing process into a kind of game, it becomes *fun*, distracting me from what could otherwise feel burdensome.

Adding an element of play to writing can lighten the load and be a distraction from the burden writing often is. Managing such distractions is important for both getting down to work and powering through. So before you even sit down to start writing, turn off your phone and put it away, preferably in another room. Turn off the notifications on your computer. Close any irrelevant or unnecessary windows and programs. If there are too many distractions at home, use noise-cancelling headphones or go to a library or coffee shop. If you listen to music, make sure it complements rather than detracts from your work. Promise yourself that you will avoid checking email or scrolling through social media until you have done a certain amount of work or worked for a certain span of time. You might think to yourself, "It will just take a second," but the temptation of that second can kill any momentum you have built.

Distractions kill momentum. Some, however, can be productive. If you are stuck on a project, and have enough time, sleeping on it might help. Less drastically, if the writing stalls, explore productive distractions: taking a short walk, doing a few minutes of stretching, or even briefly taking on another school-related writing task. Such **productive distractions**, or distractions in a productive mode, can get your unconscious mind in gear and help you get over whatever hurdle you are facing. I find most of my own writing blocks can be overcome during an outdoor walk or treadmill jog. If you like to take walks, even short walks, to shake away the cobwebs, take a notepad with you or use the memo feature of your phone to record whatever thoughts may occur to you about the project, from your general approach to specific wording.

Managing Distractions

- Turn off your phone and leave it elsewhere.
- Turn off notifications and close irrelevant windows.
- Find a quiet place or use noise-cancelling headphones.
- Resist the urge to check your email or social media.
- Use potential distractions as delayed rewards.
- Use work as a productive distraction.

When even getting started seems too difficult, there are a few strategies I recommend. One is to force yourself into a cold start, which is easier to do if you believe that the writing you are producing will not show up in the final draft. Just the act of writing itself, especially if nothing is riding on it, can prime you for the real work and jump-start the writing process. We tend to be unable to start or continue work if it seems too heavy a burden to be worth picking up or for our capabilities and skills to handle. It may seem "impossible" to write a term paper of a few thousand words, even if we have several weeks to do it. For example, suppose you need to write a short paper of one thousand words. To some writers, this does not seem onerous. But to others, especially if they have other work to do and other commitments besides, it does.

Pro Tip: Major Writing Assignments

Break larger writing assignments into short, manageable tasks.

My best advice for getting started on and staying motivated to complete a writing task is this: break large tasks into smaller ones and small tasks into very small ones. Even if the burden of writing one thousand words seems too much for you, somehow the task of writing one hundred words—and then another hundred, and then another, and so on, up to one thousand—becomes comparatively easy. To be clear, the idea is not to sit down and think that you should write one hundred words ten times. Rather, you should *not* sit down to write an essay. You should sit down instead to write what feels like an easily manageable number of words, then maybe repeat the process, and so on. Even if you are already halfway through a thousand-word assignment, think only of writing the next manageable slice of text. Let me also be clear that I do not necessarily recommend dividing writing tasks by 10 or into hundred-word chunks. Maybe a reasonable goal in a certain case is a full paragraph, or a page, or even—if you find it hard to get started—a single opening sentence. If you cannot seem to start, but then manage to meet an achievable goal, like writing that first line, you could end up having a productive writing session.

Motivating Yourself to Write

- *Work with rewards*: deferred gratification, creative procrastination
- *Make it fun*: the challenge, potential for play and game
- *Manage distractions*: minimizing the negative, exploring the positive
- *Jump-start when needed*: material to be discarded, manageable tasks

The Bookends: Prewriting and Editing

Not all writing requires what I am referring to here as bookending. In formal writing, however, prewriting before the draft itself and editing after the text is written are almost always extremely helpful if not indispensable adjuncts to the writing process. **Prewriting** is not a matter of what you do before you start to write but rather the writing that you do *prior* to the writing of your draft. We may understand it either as something that comes prior to the writing process (as in *pre*-writing) or, alternatively, as the first part of that formal writing process. The more familiar concept of editing involves double-checking, revising, polishing, and proofreading your work once it has already been drafted. Some writers prefer to edit as they go, which is perfectly fine, but in such cases, too, a post-draft review is usually helpful.

In most good writing, there will be an identifiable purpose, a *main* point. This is certainly true in formal writing, especially in academic and professional contexts. In such writing there is, or should be, a clear purpose or purposes that the writer aims to achieve. You might think that the goal of writing an essay is to earn a grade on the way to getting a course credit. This is important, of course. But the main purpose of a piece of formal writing should be indicated in the writing itself. For instance, in your term paper, the purpose of the writing is not to get a grade—even if this is the ultimate objective in your mind. No, the purpose of an academic essay is, typically, to support your thesis statement. If a work seems to have more than one objective, the main purpose is to establish those various objectives or, in other words, the *set* of subsidiary goals.

The purpose of prewriting is to help you discover your main point, if you have not yet discovered it. You may know what your main point is even before prewriting, but sometimes you discover it only by sifting through lots of research material first and then trying things out on paper. Besides this, prewriting is meant to help you develop support for the main point, to organize and connect the support. This sets you up to begin the drafting process. One type of prewriting is simply stating, in a preliminary way, what you think the main point is or should be, for instance, the thesis of your essay, which often should be revised as you proceed. Other types of prewriting include listing various points to include and brainstorming ideas and relations among them, and of course outlining. You can also try **freewriting** (Langan et al., 2009), in which you just write automatically, without any preconceived notions, to *discover* or work out your main point or how the different pockets of material interrelate.[4]

Prewriting Tactics

- Formulate your main point, subject to revision, when you can.
- List various ideas and source materials to develop your main point.
- Use brainstorming and freewriting to discover connections.
- Organize your interrelated material into an outline.

After drafting, you may be tempted to breathe a sigh of relief, feeling that the task is done. The temptation is natural and strong. But, even in the relatively minor matter of emailing a professor, proofreading your work and editing where appropriate are wise practices, and they are more than valuable. They are all but indispensable in formal writing contexts, including emailing your professor, which counts alongside lab reports and term papers. Too much is riding on your writing for you not to double-check your work. Where possible, set aside a text to look at later with fresh eyes. After drafting an essay—which you might have wisely planned to finish days before the due date—review your first draft after a good night's sleep. If you have less time to work with, you can always just take a break and then check your work. In this way, you can catch sloppy and potentially embarrassing errors, not to mention thinking up better ways to express your thoughts.[5] Even great writers tend to find their work improves significantly when they double-check and edit their work.

Pro Tip: Prewriting and Editing

Place just as much emphasis on prewriting and editing as on drafting.

PRINCIPLES OF COMPOSITION: RULES OR GUIDELINES?

There are principles to keep in mind when writing in formal contexts, some that apply to formal writing generally and others that apply to academic writing specifically. In general, early on in a piece of formal writing, you should identify your topic and purpose: what you are writing about and what you are trying to do. You should also do your best to ensure the accuracy of the information you present, stay relevant to the subject and purpose, be consistent, and order your material logically. Another good piece of advice with formal writing is to allow yourself extra time for prewriting, drafting, and revising. In academic writing, it is even more important to write as clearly and precisely as you can, to define key terms—using technical rather than dictionary definitions—and to give evidence for conclusions rather than simply stating claims.

More specific advice comes from instructors and style guides. You might recall English teachers giving you advice such as "Never use the passive voice" or, if they were being cheeky or subversive, "The passive voice should never be used."[6] Style guides usually advocate similar principles of composition. Students tend to err at one of the two extremes, either ignoring such advice altogether or devoutly following it as a set of unbreakable rules. Instructors often claim that such rules are exceptionless and should *never* be broken. But many such principles do have exceptions, not only in informal but also in formal writing contexts. In some cases, the advice is so good as to be close to an ironclad rule, yet in other cases, we find what are, at best, loose guidelines that simply do not apply in many cases.

This is certainly true of many style guides modelled after the classic *The Elements of Style* by Strunk and White (2000).[7] As you would expect, such general style guides are useful in many different contexts, but they can seem dated in some respects and less applicable to the particular writing tasks faced by kinesiology students. In what follows, we will examine different pieces of advice from Strunk and White, some of which you should take as virtually ironclad rules and others that you should take with a grain of salt, that is, as rough guidelines that often will not apply to writing in kinesiology. I will not discuss all the principles of composition in *The Elements of Style*, since my purpose is to note certain key pieces of advice while acknowledging that how strictly these principles should be taken varies considerably. Again, some are strict(ish) rules, others rough guidelines.

Strict(ish) Rules of Composition

Presenting different principles of composition as being of equal importance is useful for purposes of coherence, but is potentially misleading. Students must use such advice with discernment and good judgment.

One piece of advice from Strunk and White that should be followed most of the time, if not always,[8] is to "*make the paragraph the unit of composition*" (2000, p. 15). Whether you are writing a term paper, a lab report, or even an email to a professor, think of the text as breaking down into units, each one a paragraph. The overall design of the text breaks down into these units, and these units have their own internal structure as well as being structured in order and individual role. The length of any paragraph will depend on the function of the paragraph within the whole text and what kind of development or support may be needed to realize that purpose. For instance, the opening paragraph of an email might consist of a simple sentence in which you announce the purpose of the email (see "Email Tips" box later in this chapter). The second paragraph, by contrast, could be quite extensive, running hundreds of words.

Another principle that should be followed virtually all the time is to "*express coordinate ideas in similar form*" (Strunk & White, 2000, p. 26). The idea here is to use **parallel structure**, or similar grammar for a series of different elements, when it is possible to do so. The following example will illustrate:

- Attendance is mandatory for professors and those who are students. ✗
- Attendance is mandatory for professors and students. ✓
- Attendance is mandatory for those who are professors and those who are students. ✓

You may notice that, of the two recommended sentences, the second is somewhat wordy, and so will often be inferior to the first. Even so, the wordy suggestion is a marked improvement on the original and may also, in some contexts, be preferable to the streamlined option. If, for instance, you are discussing various groups who contribute to a university, *those who are* is an implicit reminder of the main topic: *those* (contributors) *who are*. It may even be possible to find outlier cases where it makes sense *not* to use such parallel structures, but they will be extremely rare.

A third piece of virtually ironclad advice from Strunk and White is to "*keep related words together*" (2000, p. 28). Again, an example will illustrate:

- The dancer wobbled doing a pirouette during the performance stage left. ✗
- The dancer wobbled doing a pirouette stage left during the performance. ✓

Because the phrase "stage left" indicates where the pirouette took place, it is confusing to have it appear at the end of the sentence, far from the description of the action itself. This separation is as clumsy as the pirouette described.

Again, avoid interpreting all such advice as necessarily unexceptionable. Although some of these rules may be firm, others are less so, and you will need to interpret their use appropriately.

Rough Guidelines for Composition

You might note the irony in me telling you *not* to simply accept other people's pronouncements, because that itself is a pronouncement. True enough. But there is a difference in that I am giving advice *about* advice, in other words **meta-advice**. With that in mind, let us turn to Strunk and White examples that should be interpreted as no more than rough guidelines, sometimes inapplicable to writing in kinesiology.

Strunk and White are not alone in recommending that we "*use the active voice*" (2000, p. 18). This advice is common. In the **active voice**, a sentence focuses on the agent of the action described. By contrast, in the **passive voice**, the sentence is about something other than the agent of the action described. Take the following examples:

- The winning goal was scored by Christine Sinclair.
- Christine Sinclair scored the winning goal.

In these examples, the first illustrates the passive voice. The subject of the sentence, "the winning goal," is not the performer of the action: Sinclair is. In the second sentence, the active voice is used; the subject of the sentence, "Christine Sinclair," refers to the performer of the action, Sinclair herself. If we take too strictly the advice to use the active voice, we will always prefer the second formulation to the first.

However, there are many cases where the passive voice is a preferable, even unavoidable, option.[9] For instance, in most scientific writing, you are supposed to avoid the first-person *I* and *we*. Consider which of the following descriptions of experimental trials is most suitable to include in a lab report:

- We performed trials on a biweekly basis. ☒
- Experimenters performed trials on a biweekly basis. ☒
- Trials were performed on a biweekly basis. ☑

In this case, the third sentence, decidedly in the passive voice, is most suitable for lab reports and other scientific writing.[10] The same holds for paragraphs in which you are talking about topics themselves, which, as abstractions, cannot perform actions. If you are writing about flexibility, for instance, it may be better to write, "Flexibility has been defined as …" rather than "So-and-so defines flexibility as.…" Even in the example above about Sinclair's winning goal, if the topic of the paragraph is the goal itself or another aspect of the game, it might make sense to use the passive voice. You can achieve parallel structure by writing "The first goal was scored by …" followed by "The second goal was scored by …" and so on.

You should also take with a grain of salt advice such as "*put statements in positive form*" (Strunk & White, 2000, p. 19). The idea here is to avoid weak or lazy uses of the word *not*, preferring, for instance, "The captain's advice was useless" to "The captain's advice was not useful." Taking this advice too strictly, however, you may forgo the emphatic and subtle possibilities of using negative forms. If we are entertaining someone else's claim that the captain's advice was useful, the *not* formulation could be a more balanced and emphatic option: "Some players thought it useful, but the captain's advice was *not* useful." As for subtle uses, *not* sometimes forms double negatives that fail to collapse into simple positives. A lukewarm response to the captain's advice may leave you neither impressed nor unimpressed but rather *not un*impressed. In short, *not* can add shades of emphasis and meaning beyond mere positive forms.

Finally, here is a counterintuitive example. Strunk and White advise us to "*omit needless words*" (2000, p. 23), which is sensible-seeming advice going back to George Orwell at least. The funny thing about such advice is not that the advisors themselves sometimes ignore it, but rather that Strunk and White fail to acknowledge that the principle is only one among several competitors having reasonable claims. For instance, Strunk and White (2000) suggest replacing a wordy phrase like "there is no doubt but that" with "no doubt" or "doubtless" (p. 23). True, many students could clean up their writing by avoiding wordy formulations. However, other principles, including considerations of style, may override this prescription. Using the wordier formulation may better fit the rhythm of your prose or allow for parallel structure or other stylistic qualities that, all things considered, are preferable to the terser formulation. So such "needless" words are needless for making your point but not necessarily needless for making it effectively.

Interpreting Principles of Composition

- Take some principles as strict(ish) rules, for instance
  - See paragraphs as the units of composition
  - Use similar forms or parallel language if possible
  - Keep together content-related words and phrases
- Take some principles as rough guidelines, for instance
  - Use the active voice
  - Put statements in positive form
  - Omit needless words

THE IMPORTANCE OF STRUCTURE

Formal writing is structured, sometimes explicitly, sometimes implicitly. Such constraints often feel restrictive, but they can actually make writing easier. When writing a standard resumé, for instance, you begin with your name and contact information, followed by itemized education, a chronological list of work experience, a description of related skills, and perhaps a list of hobbies and references. Having that structure laid out for you makes the task of writing the resumé easy because, in effect, you are plugging information into a template. The structure is laid out for you.

In some kinds of formal writing, however, the appropriate structure is less obvious. Even where such structure is left implicit, you should keep it in mind as a guide. In high school, you may have been required to write five-paragraph essays. At the college and university level, however, you will have more freedom. Even so, formal essays ought to be structured, beginning with an introduction, ending with the conclusion, and containing a main body aimed at establishing the thesis statement given in the introduction. To conclude this chapter, I will consider examples of formal writing where the appropriate structure is explicit (lab reports) and where the appropriate structure, though implicit, should still be followed (formal emails).

Example of Explicit Structure: Lab Reports

The lab report is one example of formal writing where the proper structure is explicitly laid out for you. Indeed, as with resumés, the standard lab report structure is reflected in various sections, each with its own proper heading. This format can vary somewhat across kinesiological subdisciplines and among instructors, and sometimes you may be asked to do less formal lab reporting, with fewer strictures on elements and format. Most of your lab reports should, however, follow something close to this format:

- *Introduction*: the topic to be addressed, in-text citations of previous related research, the purpose of the study, and the hypothesis tested. A hypothesis is, roughly, an educated guess about how the independent variable(s) will affect the dependent variable(s).[11]

- *Method*: subject/sample characteristics and selection, the equipment and materials used, the data collection procedure.
- *Results*: a description of your findings, often accompanied by detailed representations (in tables or figures) of the analyzed data.
- *Discussion*: your interpretation of whether and how the results bear on the hypothesis and compare with other relevant research findings.
- *References*: bibliography including full citation data for all and only sources cited in-text, where required data and correct order of entries depend on the documentation style.[12]

Example of Implicit Structure: Formal Emails

Where an explicit structure is not provided, you should think of any piece of formal writing as being implicitly structured. You should regard even something as simple as an email to a professor as having a proper beginning, middle, and end.[13] Even an informal grocery list is best structured in order of your typical route through the grocery store. Since our concern is with formal writing, however, we will look at how academic and professional emails should be structured.

Pro Tip: Formal Emails

Write formal emails as you would write formal letters.

In short, you should write formal emails as you would formal letters, even when you are emailing a professor you know well.[14] You may think the recipient should know, as you are in fact emailing them, that you mean to do so, and that beginning with a letter-style greeting is redundant—and in a way it *is*. Nonetheless, you should begin with an appropriately formal salutation, such as "Good afternoon, Dr. Holt," followed by a colon or comma. Since you should try to err on the side of formality, the exception to this rule is when the professor has told you directly that they prefer to be addressed more informally. Likewise, you should end with a suitable sign-off as in a formal letter, including your name, even though the recipient knows, having received the email from your account, who sent it.

Also in line with formal letter writing, you should announce in your first line the subject and the purpose of your email, as in, "I write to ask for an extension on the upcoming assignment in our communication class." Limit the body of your email to giving enough clear, relevant detail to serve the announced purpose. In this hypothetical case, you might briefly describe your grounds for the request: perhaps you are ill and are providing a doctor's note to that effect.[15]

There are certain aspects of formal email writing that differ from formal letter writing. In both cases, of course, it makes sense to double-check your writing before sending it. But it feels more natural to do so when you sign, fold, and stuff a letter before sealing the envelope. Since it is comparatively easy to click "Send" prematurely, remembering to double-check is that much more important for formal emails. Emails are less like letters and more like memos, in part because of the subject line, which students often neglect to use or use well. Because professors receive a great number of emails every day, it is crucial for you to use a subject line with language that indicates that your email concerns them specifically and is not spam or about something general or frivolous. Sticking with the above hypothetical, your subject line could read, "Extension request for term paper, COMM 1013." This is much more likely to receive prompt professorial attention than a blank subject line or a vague "Extension." It is also easy to forget to attach an attachment after you have written your email. How often have you received two emails, sent minutes apart, the first asking you to check an attachment that is not there and the second an apologetic "Oops"? To reduce the chances of creating an impression of carelessness, attach attachments, if any, *before* you begin composing the email itself.

Formal Email Tips (Step by Step)

1. Attach attachments, if any, before composing the email.
2. Fill in the subject line with precise subject-specific phrasing.
3. Open with a salutation, including the recipient's correct name and title.
4. Announce the purpose(s) of your email in the first sentence.
5. Craft the structure and length to include all and only relevant details.
6. Sign off appropriately, including your name.
7. Double-check everything before clicking "Send."

Let us end the chapter with a hypothetical email from a student to a professor, requesting an extension on an upcoming assignment. Although this is not exactly a template you should use in every case, it does illustrate many of the important virtues of formal emails:

Email Example

From: john.smith@fictionalu.ca

To: jane.doe@fictionalu.ca

Subject: Request for extension on KINE 101 essay

Sent: 3:15 pm, Wed 10/10/2022

Attachment(s): Note from Fictional U campus clinic (PDF)

Dear Prof. Doe:

I am a student in your KINE 101 class, and I write to request an extension on the essay assignment due this Friday.

My reason for making this request is that I have been seriously ill since late last week and so have been unable to do much of the work I had planned. Following the extension policy given on your class syllabus, I have attached a doctor's note from the campus clinic to document this issue.

Thank you for considering this request.

Sincerely,

John Smith

Here are several noteworthy facets of John Smith's email to Professor Doe: it includes a specific, relevant subject line; it is sent two days before the deadline rather than the night before; it greets the professor and uses her correct name and title; it announces the purpose of the email in the first line; it gives sufficient detail and justification to support that purpose; it acknowledges the professor's applicable policy; it includes an appropriate sign-off; it maintains a respectful tone throughout; and it is free of errors in spelling or grammar. Overall, it is a successful email worth emulating. You should keep such considerations in mind for all your formal writing tasks.

BRIEF RECAP

In this chapter, we explored aspects of composition:

- What composition is
- The writing process and how to think about it
- Principles of composition: rules or guidelines?
- The importance of structure

CHAPTER 3

Communicating Responsibly

CHAPTER OVERVIEW

Communicating irresponsibly can lead to all kinds of unexpected difficulties. In this chapter, we focus on what it means to communicate responsibly. We begin by introducing the concept of communication ethics, followed by a brief look at some of the pitfalls of online communication. We discuss the important issue of inclusive language and writers' responsibilities as well as the general importance of respect for the truth and other people. Finally, we examine how subtleties of language may unintentionally suggest unfortunate and unintended interpretations of what we say and write.

COMMUNICATION ETHICS

Communication ethics matters. Although both ethics and morality are concerned with principles of right and wrong, morality concerns what is required to be a good person and to perform good deeds in general. Ethics is narrower. It also concerns principles of right and wrong, what is and is not permissible, but only in certain contexts. These contexts are often, though not always, formal and involve specific roles occupied by people who are expected to live up to the principles involved. Not all such roles are professional, although many are: medical ethics for doctors, business ethics for businesspeople, and so on. There is also sports ethics for athletes, which applies irrespective of professional or amateur status. These role-specific principles and codes of conduct apply general moral principles to specific situations (e.g., as a doctor, in business, on the field, and so on) and tend to be more demanding. As a communicator you, too, are adopting a certain role and should follow ethical principles pertaining to that role and its context accordingly. We all should try to be ethical communicators in general, of course, but also specifically in formal academic and professional contexts.

Even in terms of self-interest, you should care about being a responsible communicator. If you communicate *ir*responsibly, either deliberately or through carelessness, the consequences can be severe. At best, an irresponsible remark or careless social media post could lead to social or professional embarrassment. At worst, it could lead to formal reprimands, fines, or losing your job. Both ethics and self-interest should move you to communicate responsibly.

As kinesiology scholars, you do not have anything like a standard "10 Commandments of Communication Ethics" to guide you.[1] Nonetheless, you should be aware of and take to heart something like the Golden Rule. As Williams and Colomb (2012) articulate their First Principle of Ethical Writing, "we are ethical writers when we would willingly put ourselves in the place of our readers and experience what they do as they read what we've written" (p. 142) or simply "write to others as you would have others write to you."[2] Whereas Williams and Colomb frame this as a principle of *style* with ethical implications for writing, I would urge that it is an *ethical* principle with style implications and is applicable to formal communication in general. In other words, I suggest that you, as kinesiology students, communicate with others as you would have them communicate with you. Note that this formulation applies to, for instance, giving class presentations and even chatting casually with fellow classmates, as well as engaging in more formal exchanges with professors and administrators. In all such interactions, you should conduct yourself in an appropriately respectful manner.

The Golden Rule of Communication

Communicate with others as you would have them communicate with you.

ON BEING ONLINE

Though we are obviously most concerned here with formal academic contexts, the Golden Rule of Communication applies to informal contexts as well, so a few remarks about communicating online are in order. A paradigm case of unethical communication online is the internet troll. One would find it hard to image trolls themselves wanting to be trolled. Skewing negative, however, is hardly the only problem. Many social media posters flirt with dishonesty in how they present themselves, cherry-picking content that shows them at their absolute best and using software to artificially enhance their images. Even honest self-presentation online may have unfortunate consequences. We have all heard about how careless social media posts, even apparently innocuous posts of people enjoying themselves at parties, can hurt people professionally. Informal and formal communication contexts are simply not as completely distinct as many of us would like.

Even getting too caught up in the consumption of online content has its problems. A theorist named Walter Benjamin (2002) identified an unfortunate consequence of mechanically reproducing images. In his view, mechanically reproducing images of important artifacts such as the *Mona Lisa* is good because it gives many people access to those artifacts, which they might not otherwise have. But this also has a downside. It often detracts from the special significance a thing itself has in real life—which Benjamin describes as an "**aura**." This applies no less to digital reproduction of images online (Davis, 2002). We all know what the *Mona Lisa* looks like because we have seen photographs of it in books and online. By contrast, relatively few of us have travelled to Paris to see it "in the paint" behind bulletproof glass in the Louvre. Many of us who have made the trip are disappointed by the experience of the actual painting. Its aura has been diminished. And not only artifacts can lose their aura. People can as well. If you are overexposed to someone via social media, that can diminish the significance of their real-life presence for you. This loss of aura can happen with in-person overexposure as well. However, the point is that overexposure to online experiences via social media may diminish the special meaning of people and things in the real world.

A related problem pertains to what another theorist, Jean Baudrillard, referred to as the **hyperreal**.[3] In some cases, representational images of things, or simulations, become more important than their corresponding reality. Imagine caring less about whether you have a good life and more about whether your life presents well on social media. When representations become more important than reality, the result can be psychologically dysfunctional and morally problematic.[4] According to Baudrillard (1994), these simulations can ultimately not just become more important than reality but can become reality itself, as if the "real reality" has been somehow lost in the shuffle. What it means for a simulation to "become real" is somewhat unclear, but it should give you pause about spending too much time and energy online. Ethical concerns in communication contexts, both online and off, should include not just other people but also you yourself. The point, as with the loss of aura through overexposure to images, is that we should be careful about losing our grasp of a real world we care about by becoming caught up in a virtual hyperreality online.

INCLUSIVE LANGUAGE

Communicating ethically limits not only what we use language to do but also what language we use to do it. In formal contexts especially, it is incumbent on us to use **inclusive language**. Such language is ethical because it recognizes how language may have, and historically has had, the effect of reinforcing the marginalization of certain, often disadvantaged, groups of people. Recognizing this, inclusive language seeks to diagnose such cultural trends and correct them. Whereas biased or prejudiced language excludes people, inclusive language is meant to do the opposite. Ideally, we should be sensitive to the social issues that continue to motivate inclusive language use. Yet even if you find yourself unmoved by these concerns, it is still vitally important to scrupulously follow the conventions of ethical language use within scholarly and professional disciplines. This may seem arbitrary, as if we are trying to keep up with ever-changing fashions. In a way this is true. But these fashions or ways of speaking are principled, not arbitrary. So whether we agree with such principles or not, we are responsible for keeping our language choices in step with whatever conventions are in effect in academic and professional contexts. We might intend no harm and harbour no ill will by failing to adopt inclusive language, yet we may still cause offence and run afoul of guidelines we should care about and are expected to follow.

In addition to respecting the ideals of sensitivity and following formal conventions, we should bear in mind that inclusive language means putting aside our own plausibly biased linguistic habits and preferences and prioritizing instead others' terms of self-reference. And such terms of reference may vary depending on whether the speaker is or is not recognized as part of the group in question.

Gender-Neutral Language

Perhaps the most obvious example of inclusive language is the use of gender-neutral terms when appropriate. Use of literally gender-neutral terms, such as the pronoun *their*, can be convenient in many contexts. For instance, the bromide "To each his own" can be better rephrased as "To each *their* own." In some cases, there will be obvious gender-neutral terms to substitute for gendered ones, as with *server* instead of *waitress*. In other cases, the key might simply be avoiding gender specificity when there is no reason to be gender-specific. There is often little reason, for instance, to describe someone as a *female* professor, or doctor, or coach, rather than simply a professor, doctor, or coach.[5] Being gender-inclusive can sometimes be as simple as varying the details of your examples. If you are exploring a hypothetical scenario, for instance, consider naming your imagined person, say, Letitia rather than Andrew, especially when they perform a traditionally male-dominated role. Using gendered names *against* stereotype implies an open, progressive attitude and helps undermine those stereotypes. If you are unaware of an actual person's pronouns, rephrase statements with gendered pronouns to avoid unnecessary commitment. Refer to the person by name ("Pat gave a great presentation") or choose a gender-neutral description ("That speaker did a great job").

Pro Tip: Navigating Uncertainty about Pronouns

Where pronouns are in doubt, use names or appropriate gender-neutral descriptions.

Avoiding gendered language is not always easy. It is common in sport contexts, and because of a lack of gender-neutral equivalents, or even for convenience, the formal use of gendered terms may be appropriate. Consider Leslie Howe's discussion (2013):

> Some will find the terms "gamesmanship" and "sportsmanship" to be gender biased. As a woman who has and continues to participate in traditionally male sports on male, female, and mixed teams, I am hardly indifferent to gender bias in the cultural definition of sport and its language.... However, although there might be some important deconstruction to be done regarding historical and normative assumptions behind the apparently masculine bias of these terms, for the purposes of the present discussion I shall employ them as is for the fairly practical reason that they correspond to the common usage of sports participants (rightly or wrongly) and because I here take the "man" in "gamesmanship" to have the same gender-indifferent marker quality that it has in a directive such as "cover your man," to which an entirely appropriate response might be "I've got her." (p. 277)

Sportsmanship is a particularly interesting and difficult case. Although we could substitute the adjective *sporting* for *sportsmanship*, the connotations are not quite aligned. Likewise, although substituting *person* for *man* works in many cases, the noun *sportspersonship* seems a clumsy if well-intended substitute. The reason we don't have a good gender-neutral replacement for *sportsmanship* is that it hasn't been created. I have proposed the neologism *sportitude*[6] for this role (Holt, 2014, p. 297). So far it has not caught on, but I hope it does. In any case, you might be able to come up with something better.

The *Sportsmanship* Challenge

Try to come up with a creative gender-neutral substitute for *sportsmanship.*

Other Types of Inclusivity

There are several other types of inclusivity in language use that you should be aware of, and we will discuss them briefly. In particular, we will now cover sexuality and gender identity, people with disabilities, and race and ethnicity. Again, it will behoove us to be sensitive to others' needs and identities when we refer to them. This is not a matter of catering to people's whims. We are not talking about frivolous or arbitrary preferences here, but ethically significant choices. We can see how such considerations play out in particular types of designations.

When it comes to people with disabilities, the current general convention is to emphasize the person rather than the disability. Hence phrases such as "person who is blind" or "person with visual impairment" are preferable to placing *blind* or *visually impaired* before *person*. Still, in some cases people might prefer to highlight their condition as part of their identity, preferring to be called, say, "a deaf person" rather than "a person who is deaf." Case by case, you can depart from conventional assumptions out of such preferential respect. But the default assumption should be to err on the side foregrounding the personhood of those with such conditions.

With respect to sexual orientation and gender identity, self-referential preferences come to the fore. Members of the LGBTQ+ community sometimes permit internally appropriate terms for one another that are inappropriate for use by those outside those communities, even outsiders who consider themselves allies. Likewise, *sexual orientation* is the preferred term rather than *sexual preference* because the latter suggests a choice and, for many, something less closely connected than is fitting to a sense of identity.[7] Terms of self-reference are even more salient when we consider gender identity, as it is becoming increasingly common to use pronouns that reflect an individual's gender identity—often in contrast to cisgender categorization. The conventions governing appropriately respectful usage here are in a state of flux.

Appropriate ethnic and racial terms are also important, of course. In Canada, for instance, the broad term *Indigenous* is preferred, in many contexts, to refer to people descended from the land's original inhabitants, although in other contexts, more specific terms, such as *First Nations*, *Inuit*, or *Métis*, are more appropriate, depending on specificity of reference. Appropriate specificity is key, even to the point of mentioning a particular person's First Nation. Here it is especially important to make sure you are using the correct terms and spelling. For instance, in Nova Scotia we used to use the word *Micmac* indifferently as a noun, singular or plural, or adjective to refer to people from regional First Nations communities. Now, however, the conventional spelling for the plural specifically is *Mi'kmaq*, whereas the singular and adjective form is *Mi'kmaw* (Office of Aboriginal Affairs, n.d.).[8] Keeping up with these linguistic evolutions is challenging but also vital, and there are usually resources to help, whether at your library or on a government website.

Should Indigenous Terms Be Italicized?

It makes sense to italicize Indigenous terms such as *Mi'kmaq* to both indicate and respect that they are not English words. However, doing so suggests that we are treating the terms as if they are *foreign*. Indigenous languages in the Americas have a better claim to *domesticity* than do English, French, and Spanish. Should we italicize colonial languages, or does avoiding italics for Indigenous terms both honour their domestic origin and promote awareness of appropriate use in the culture at large?

Inclusive Language

1. Avoid racial and ethnic slurs and generally offensive language.
2. Choose gender-neutral terms where appropriate and possible.
3. Respect other people's terms of self-reference.
4. Use current terminology for communities and their members.

WRITERS' RESPONSIBILITIES

You have certain responsibilities to live up to, both as a writer and as a communicator generally. Some of these responsibilities are obvious, others perhaps not so. In both cases, it will be helpful to address them explicitly so you will be more likely to keep them in mind when you communicate in formal settings. Respectful communication has implications for how we speak to others, even how we listen, but for convenience, I shall discuss these as writers' rather than communicators' responsibilities. There are four types that you should keep firmly in mind when writing formally. These responsibilities can be distinguished by their objects: by whom or what you are responsible *to* as a writer.

First, as a writer in formal contexts, you are responsible to *readers*: you owe them. One thing you owe them is honesty. In your writing, avoid trying to deceive them or using tricks to persuade them of something you do not believe yourself—unless you are asked to do so as an exercise. One especially egregious form of deception is plagiarism: trying to pass off someone else's work as your own. But this is not all you owe readers. You also owe them clarity. In other words, you should make what you are trying to say as accessible as you can. Even if you are sure your professor knows what you mean, the more accessibly you write, the better they will be able to see that *you* know too. If you have to use a technical term, define it. If you can choose simpler language instead, do so. There are limits on how accessible you can make your content, particularly when the material is difficult. Nonetheless, you are obligated to do your best. This is not everything that writers owe to readers, but the responsibilities mentioned are among the more important. Respect for a reader sometimes means knowing something about who they are, along with their specific needs and tastes, which becomes more challenging as your readership becomes wider and more diverse.

Second, as a writer, you are also responsible to your *subject*. You should try to treat your subject as fairly as you can. This complements your responsibility to be honest with readers. You should present your material in a fair light, trying your best to avoid misrepresenting it. One part of this process is presenting the subject in a way that appropriately conveys its degree of difficulty. If, for instance, you flippantly describe kinesiology as "phys ed with scientific pretensions," that is unfair to the discipline. You should not water down or otherwise distort content, even for the sake of accessibility.

Note a possible tension here: on the one hand, you are supposed to write clearly for your reader, but on the other hand, you are not supposed to present difficult material too simplistically. There will be cases, especially in more advanced research, where these are competing objectives, where accessibility and appropriate difficulty are at odds. The call to clarity of your text may conflict with the call for clear representation of difficult material. Sometimes you might need to sacrifice clarity that could potentially be misleading for the sake of appropriately representing difficult material. Kinesiology is not phys ed with scientific pretensions but rather the study of human movement and its role in our lives and culture as treated from a variety of disciplinary perspectives ranging from physiology to philosophy. But there is no general rule for resolving these tensions, and you must resolve them, therefore, on a case-by-case basis. The more obvious problem, however, is indulging in deliberately or unnecessarily obscure writing, in effect hiding certain features of your material. Perhaps you are trying to sneakily put one over on your reader, for instance, making the content seem more challenging or significant than it really is. In a word, don't.

Third, in formal contexts especially, you do not owe something only to the subject and the information; you also owe something to your *sources*. As with the call to honesty, here too the issue of plagiarism comes into play. You must properly acknowledge sources used in your research to avoid falling into unethical behaviour, violating standards of academic integrity. But acknowledgement of sources is only part of the ethical requirements. You also owe these sources fidelity and charity. You owe them fidelity in the sense that you are supposed to represent them as accurately as possible, as with your subject matter.[9] Misrepresenting source material, whether through malice or carelessness, means failing to live up to this responsibility.

Owing sources charity means interpreting texts and writers *charitably*, attributing to the text and to the writer's intention as plausible a reading as possible given the textual data at hand. This is sometimes called the **principle of charity**. If the author of a text claims that there might be reason to doubt the legitimacy of bans on performance-enhancing drugs in *some* sports, it would be an uncharitable interpretation to present them as being steroid-friendly.

The Principle of Charity

In interpreting source material, give the author(s) and the text the benefit of the doubt.

It may seem that all the bases are now covered. As a writer, you are responsible to readers, to the subject matter, and to whatever sources you may use. Is that all? There is one more object of responsibility: you yourself. As a writer, you owe yourself certain things. You are responsible for your writing, and your writing is a representation of yourself, no less than any private journal entry or social media post.

This applies even if you do not want it to serve that purpose, even if you don't really care about your writing at all. Given this, with any piece of writing, you owe it to yourself to put in a good deal of effort to get it right. Ideally, you should care about the writing as a representation of yourself. You should care about how it reflects on you as a student, as a thinker, and as a person. In the same way that people may care about their appearance or take some pride in personal grooming, you should take pride in the writing you produce in formal contexts. When it comes to writing, you owe it to yourself to make an effort, to *care*. Putting your best foot forward when it counts is less about how you look and more about how you come across to others. Like it or not, that includes how your writing comes across.

Writers' Responsibilities

1. To readers: honesty and clarity
2. To subject: fairness and difficulty
3. To sources: fidelity and charity
4. To themselves: effort and care

WITH RESPECT

Underlying the call to be honest with readers is an attitude of respect. Honesty is about respecting more than other people, those with whom you are honest. It is also about being honest with yourself, a matter of self-respect. Beyond this, the call to honesty implies a respect for the truth. The importance of respecting other people and the truth alike in formal communication cannot be overstressed. In this section, I will briefly discuss several issues involving respect for other people as well as respect for the truth.

We tend to think of being honest with other people and ourselves as just telling the truth. This is oversimplified, however. If you lie to another person, that is not necessarily a matter of failing to tell the truth, but rather a matter of trying to deceive the other person, of trying to lead them away from what you *think* the truth is. Honesty is a matter of expressing what you *think* the truth is, not necessarily what the truth is in fact. This is because we are finite and fallible beings and do not always know what the truth is, even when we believe we do.

Truth is a property of sentences, specifically assertions. Assertions or statements are the sentences that purport to say something about how the world is. If I say, "The gymnast is on the mat," this statement is true if and only if as a matter of fact the gymnast really is on the mat. A true sentence is one that corresponds to a fact, a state of affairs. Likewise, if I say, "$E = mc^2$," this is also true because it corresponds to a fact, though a rather different kind of fact, a general physical law rather than a

Figure 3.1. Truth as Correspondence
Source: Jason Holt

Figure 3.2. Falsity as Failed Correspondence
Source: Jason Holt

particular fact available to visual inspection. Another way of putting the point is to say that a sentence is true if and only if the world is as the sentence says it is. False assertions are, by contrast, those where the world is *not* as the sentence says it is.

How we know that $E = mc^2$ differs from how we know, if we do, that the gymnast is on the mat, because these are different kinds of facts.[10] Ordinary visual perception can tell us how gymnasts and mats are spatially related, but knowing that $E = mc^2$ is true depends on the empirical and conceptual expertise of physicists. In both types of case, though, correspondence with a fact makes a sentence true.[11] Respecting truth often involves stating it as far as you know it—what you believe to be the case given the evidence available to you, whether it comes in the form of your own sensory perception or someone else's experience or expertise.

Formal discourse requires that we respect the truth in this way and that we likewise respect other people accordingly. If we try to deceive others, leading them away from what we believe is the truth, that is a **lie**. In lying, we use language to deceive other people for illicit purposes and so are also using people rather than respecting either truth or other people as we should. This is especially important in formal contexts both professional and academic, and not just for professional academics like professors, but for students as well.

One would think that there is no room for profanity in communication that aspires to be formally proper. But this is not quite true. If you are sensitive to off-colour or modestly profane words, brace yourself. Being honest is not just a matter of avoiding lying, but also avoiding other morally questionable types of language use as well. One provocative example is discussed in a pocket of research on—brace for it—**bullshit**. It might seem inappropriate for me to type the word out like that. The word is particularly interesting in this context because its use in formal settings is somewhat controversial, but it is also interesting because it is a type of deceptive practice different from lying that we also have to watch out for in others' behaviour as well as our own. Because of its prevalence in our culture (Frankfurt, 2005), the term has been analyzed by researchers to the point that its use as a technical term is perhaps appropriate—even if in ordinary discourse and most formal contexts its use might continue to be considered inappropriate.

The intellectual respectability of bullshit as a topic of real scholarship owes mostly to the philosopher Harry Frankfurt, whose bestselling *On Bullshit* (2005) achieved cultural prominence. This book was originally published as a stand-alone essay in 1986. In this treatise, Frankfurt argues that the figure of the bullshitter is, morally speaking, even worse in some sense than the liar. A person being honest tries to tell you the truth. A person who is lying also cares about the truth, but tries to lead you away from it. What distinguishes bullshit, though, is that the person who is bullshitting makes an assertion *without caring about the truth at all* (Frankfurt, 2005, p. 55). So what distinguishes someone who is bullshitting, in Frankfurt's view, besides a total disregard for truth, is an exhaustive concern only with creating the right impression. The liar is, in this respect, morally superior, though still morally culpable, to the bullshitter. Along with avoiding lies, then, an ethical communicator, especially in formal contexts, needs to avoid the temptation to bullshit the reader or audience.

Discussion Question

Is it appropriate for me to have typed out the word *bullshit* rather than *b*******
or *BS*? If not, why not? Is it because, as a swear word, the term is profane and thus inappropriate in a formal context? If it is appropriate, is this because academics and scholars have discussed the phenomenon and somehow legitimized the term as a topic of cultural significance?

A related phenomenon you should be on guard against is that of **spin**, which often comes up in political discussion. The difference between spin, lying, and bullshit is as follows. With lying, you think you know the truth and are trying to lead others away from it. When you bullshit, you make claims with total disregard for the truth in order to create a certain impression in your audience, for whom the truth is elusive and a matter of little concern for you. With spin, however, the relevant facts are generally known, so there is little point trying to hide or ignore them. When you spin something, you are not trying to lead people away from the facts or create an impression on people irrespective of facts. Instead, you are trying to shape or manipulate people's interpretation of the facts, which are commonly known. The metaphor of spin, in this sense, takes for granted that *what* you are trying to put spin on is like a ball you want to curve one way or another by imparting a spin to it. What is at issue, then, is not the facts themselves but rather their interpreted significance. At least that is what the spinner is banking on. Trying to spin the facts is different from providing evidence and giving reasons for shaping people's beliefs or changing their actions. That is a matter of rational persuasion, whereas spin is a matter of trying to manipulate people. Spin involves using rhetoric and other morally questionable tactics and so must be avoided in ethical communication, as with lying and bullshit.

Since these deceptive practices are distinguished from honest statements and one another by the communicator's psychology, you cannot tell from any statement on its own whether it is, in fact, a lie, bullshit, spin, or an honest statement. If a head coach at a press conference before a big game says, "We've never been more prepared," this *could* be an honest assessment reflecting what they actually believe, but it could be a strategic lie contradicting what they actually believe, or bullshit designed to intimidate the other team, or an attempt to spin the public's perception. It may be fairly easy to police your own deceptive practices, but it can be hard to tell if others are employing them, especially if they do so skillfully.

ATTRIBUTION ETHICS

Even supposing you have taken care to communicate ethically by avoiding lies and bullshit, and have taken pains to use inclusive language and show due sensitivity, there are still other subtle biases that can, unintentionally, creep into your writing. It is not as though we can reach to the bottom of our psyches to discover and simply cleanse ourselves of all such prejudices. However, it is incumbent on us to be as self-aware as we can and be conscientious about minimizing such bias to an appreciable degree. Beyond honesty with others in formal communication using appropriate inclusive language, we should also avoid unintended and potentially biased implications.

Consider the following example. Suppose that a student, Pat, did poorly on a test and that we account for this by observing that they only put in an hour of study time, whereas students who did better on the test put in many hours. We may express this explanation in different ways, (a) and (b):

a) Pat did poorly on the test because they did not have enough study time.
b) Pat did poorly on the test because they did not plan enough study time.[12]

Students who are not particularly careful about word choice or attentive to detail in proofreading and editing might fail to notice that there is any important difference at all between formulations (a) and (b). However, the difference between saying that Pat did not *have* enough study time and saying that Pat did not *plan* enough study time is significant in terms of implications, whether or not the writer intends these implications. If we opt for (a) and write that "Pat did not have enough study time," the implication is that the poor performance was owing to the circumstances, and so Pat is not at fault. If, instead, we opt for (b) and write that "Pat did not plan enough study time," the implication is that the fault lies not with the circumstances but rather with Pat. Again, whether the writer intends such subtlety to be read into their word choice, the difference between *have* and *plan*, beyond shading meaning slightly, significantly affects the implied explanation—even in a relatively simple case. Since a reader will be unable to read your intentions and goodwill transparently through your language, attending to the language itself is no less crucial for ethics than it is for composition.

Whether such implications are unintended or express subtle bias, they may be limited to one-off cases or may reveal a more systematic and sometimes systemic type of bias or prejudice. An example of this is **attribution bias**. In explaining the relationship between people's actions, their circumstances of behaviour, and the results of those actions in context, we tend to take different slants depending on who the people are and what kind of outcomes result from their actions. People tend to praise themselves for positive outcomes and forgive themselves for negative outcomes. That is, when our own actions result in a positive outcome, we are more apt to attribute that positive outcome to ourselves as the agents of the action. But when our own actions result in a negative outcome, we are more apt to blame the circumstances in which we acted rather than ourselves, in effect marginalizing our agency. This is the so-called **self-serving bias**. Complementing the self-serving bias is what we may dub an **other-disserving bias**: a tendency to attribute positive outcomes of other people's actions to the circumstances rather than to their agency and to blame them, instead of the circumstances, for negative outcomes. Your team won because of athletic prowess, but the other team won because of good luck; your team lost because of bad luck, but the other team lost because they lacked athletic prowess.

Self-Reflective Exercise

Try to find examples in your own thinking of the self-serving or other-disserving biases. What are those examples? Do you notice such biases in others? How can we minimize such biases in ourselves and help others recognize and reduce them in their own thinking?

You can see how this subtle kind of thinking might infiltrate much of the way we see the world in an unbalanced way. To be ethical communicators, we do not simply need to be honest about what is on our minds and to use inclusive language given by ethical conventions of formal discourse. We are also called on to examine the content of our own thoughts and feelings and the potential underlying biases and peculiar distortions that may account for unintended implications of our formal communications.

BRIEF RECAP

In this chapter, we addressed different aspects of responsible communication:

- Communication ethics
- Online communication
- Inclusive language
- Writers' responsibilities
- Respecting the truth and others
- Attribution ethics

CHAPTER 4

Justifying Claims

CHAPTER OVERVIEW

The question of justification links together our understanding of knowledge, truth, and argument, as expressible in language. We begin with the traditional analysis of knowledge, which requires justification. We then look at arguments as attempts to provide justification for claims or beliefs, the distinction between strong and weak arguments, and criteria for assessing arguments. We next look at certain reasoning traps, called fallacies, to avoid and qualifying language to use in strengthening your own arguments. We finish by outlining strategies of justification to employ in critically engaging other people's views.

KNOWLEDGE

It is no great revelation that education and academia are about acquiring knowledge. Teachers are supposed to impart knowledge to students, and researchers are supposed to aim at enlarging the scope of human knowledge. At the post-secondary level, professors are supposed to both teach and research. Students at this level often find the work challenging because the material is more advanced and because the necessary demonstrations of such knowledge are much more demanding. At earlier stages of educational development, students can often succeed by brute memorization of facts and simple response techniques. At higher levels, however, such skills are still important but seldom enough for success. Professors expect students to understand material in a more sophisticated way and not just recite it, to see connections among different facts and theories, and to think critically and even creatively. In high school, a student might earn an A on an essay just by gathering facts and arranging them readably. Not so in university. In university, students will often be expected not just to research a topic but to formulate and defend their own viewpoint on the research they have conducted. In other words, university students are supposed to show their professors *how* they know what they have learned and, relatedly, *why* their views are justified. The task is not just to make claims but to *argue* for conclusions. What links knowledge and argument, then, is *justification*, and that is the topic of this chapter. Whether you are writing a quick email to a professor to ask for an extension or composing a thesis over many months, it will behoove you to always think in terms of justifying your claims.

The Analysis of Knowledge

Education and academia are about knowledge. But what is knowledge? We all have a certain idea, an imprecise grasp, of what knowledge is, but sometimes it is worth trying to more precisely specify these ordinary concepts. The analysis of knowledge underlying much current research and educational practice is surprisingly old, originating some two and a half millennia ago with the ancient Greek philosopher Plato.[1] Plato, a towering figure in intellectual history, is often considered the father of Western philosophy. On this account, to have **knowledge**, a person must take something to be true (belief), it must actually be true (truth), and the belief must be justified (**justification**): knowledge as *justified true belief*. We can represent this theory more precisely as in figure 4.1.

Knowledge is analyzed as a relation between a knowing subject (*S*) and the thing known, a proposition (*P*). You are a subject, and so am I. A **proposition** is expressed by a statement, or a sentence that can be true or false. Notice that "What's your name?" and "Close the door" are both sentences, but not of the kind that can be true or false. "The cat is on the mat" is a statement, and so is "$E = mc^2$" or "Fast-twitch muscle fibres affect sprinting ability." Propositions can be expressed by different sentences, as "It is raining" in English and "*Il pleut*" in French express the same thought. Likewise, if instead of using "The cat is on the mat" I opt for "Yoda syntax" ("On the mat the cat is"), I express the same proposition.

Statements (1), (2), and (3) are necessary and sufficient conditions, which means that if you lack any one of the three, you fail to have knowledge, and if you have all three, you have knowledge. You should think of the biconditional (↔) as an equal sign. For any given *S* and any given *P*, if you have "*S* knows that *P*," the right arrowhead takes you to the right side, which means you have (1), (2), and (3). If you likewise have (1), (2), and (3), the left arrowhead takes you to the left, which means *S* knows that *P*.

S knows that *P* ⟷ (1) *S* believes that *P*
(2) *P* is true
(3) *S* is justified in believing that *P*

Figure 4.1. Knowledge
Source: Jason Holt

As for the conditions themselves, a belief is a mental representation that the subject takes to be true (e.g., that their favourite team will play today), not one that they hope is true (e.g., that the team will win) or that they fear (e.g., that the star player will be injured). Knowledge requires belief, but not just any belief, since some beliefs are false. A true belief is one corresponding to a fact, a state of affairs in the world. A false belief is one that does not correspond to a fact. My belief that I locked my locker before leaving the gym is true if I did so and false otherwise. Knowledge implies not only belief, then, but truth. But true belief is not enough. Suppose a gambler risks all their money on red because the roulette wheel has come up black 10 times in a row and they are convinced it must hit red next time. The wheel spins, the ball falls, and they win! They certainly had a true belief that red would come up, but they did not know it.

The gambler's feeling of confidence and the fact that the belief turned out to be true does not mean they were right to have the belief in the first place or to bet on the proposition being true. Perhaps the most important condition, then, is (3), that a subject must be justified in having their belief. To say that a belief is justified is to say that it comes from a reliable source and that a subject has good reason for it. Reliable sources include our senses, memory, reason, other people's testimony, expertise, and various research methods in the sciences and humanities. That a belief is justified is no guarantee of truth, but it means that, given the information the subject has, they are right to *bet* on it being true. In our scenario, by contrast, the gambler is *wrong* to bet on their belief, even though it proves true. This is because their belief did not come from a good source but rather derived from wishful thinking, hope rather than reason.

Justification, then, is necessary for knowledge. As Plato would have understood it, for *S* to be justified in believing that *P*, *S* must be not only *aware* of the belief but also able to *articulate* it and its justification. Although today we believe that subjects may have knowledge that is partly or completely *un*conscious or *in*expressible in language, it is still vitally important to be able to access and justify many of our beliefs, especially in academic and professional contexts. In day-to-day interactions this is seldom the case. But as a student or a professional, you will often need to state your views and be able to justify them. This means that justification is not just essential in having knowledge, it is also crucial in demonstrating knowledge. The justification that you will be required to express as a student or a professional will vary according to both context and discipline. As a student, more will be expected of you on a final exam or in a term paper than on a pop quiz or in class. Less will be expected of you than would be expected of a professional or a professor. Across disciplines, too, what counts as appropriate knowledge claims and relevant evidence for those claims will vary widely, especially in kinesiology, which is interdisciplinary, running the gamut from physiology to philosophy. The kinds of knowledge (both propositions and justifications) you will be expected to learn will be markedly different in biomechanics than in a history of sport class and will be different in a lab report than on a term paper. As you progress through your program, you will be expected to understand and be able to articulate increasingly sophisticated and varied types of propositions and justifications.

ARGUMENT

Professors and other professionals will often make claims in the area of their expertise, believing others will assume they are justified in doing so. But sometimes even respected professionals are called on to back up their statements, to explain the underlying basis of their claims. University students, too, are often obliged to support their views, whether in writing essays or lab reports, giving presentations, or doing other types of schoolwork. They are asked to present their argument.

Defining *Argument*

In the relevant technical sense, an **argument** is an attempt to back up an assertion. To be more precise, an argument is a set of statements, all but one of which are intended to support the other as rationally acceptable. Arguments, then, consist of two parts: the statement one is trying to justify, the **conclusion**, and the statements meant to justify it, the **premises**. Take an example. Suppose you think that soccer is a game but are challenged, perhaps by a super-serious superfan or a professor in a philosophy of sport class, to justify this view. You might argue as in figure 4.2.

Here (P1) and (P2) are premises, the reasons you offer to justify your claim (C), the conclusion. Besides the conclusion, an argument needs at least one premise and, in principle, could have any number of them, though practical considerations invariably limit them to a manageable number. The three-statement form is particularly common, even if it takes the arguer an essay, or even an entire book, to flesh out the argument in detail.

Definition of Argument

An argument is a set of statements, all but one of which are intended to support the other as rationally acceptable.

(P1) Soccer is a sport.
(P2) Sports are athletic games. } **Premises**
Therefore,
(C) soccer is a game.} **Conclusion**

Figure 4.2. Argument

Source: Jason Holt

Arguments are not always easy to identify in ordinary language, although certain words often flag conclusions (*therefore*, *thus*, and *hence*) and others flag premises (*because* and *since* especially). We can rewrite our argument from figure 4.2 in the form of ordinary language using a variety of indicator words. We could write, for instance, "*Because* soccer is a sport and sports are athletic games, soccer is a game" or, sticking closer to the original, "Soccer is a sport. Sports are athletic games. *Therefore*, soccer is a game." (Usually, you would avoid italicizing the indicators as I have done for emphasis.) Without such indicators, however, an argument can still be present. The key is that some of the things said or written are given as reason to believe or accept something else that is said or written.

Strong versus Weak Arguments

All arguments offer reasons to accept some conclusion, but the quality of such justification can vary considerably from one argument to another. Some arguments are so compelling that the conclusions seem inescapable, whereas others are so bad that they offer not even the slightest good reason to accept their conclusions. You should avoid being persuaded by bad arguments and be open to being persuaded by good arguments. When called upon to give an argument, you should try to make your argument as strong as possible. Doing otherwise can be an intellectual or even moral failure, as I will discuss later.

The technical term to describe good arguments, those justifying their conclusions, is *cogent*. A cogent argument has two features: acceptable premises and strong reasoning. For premises to be acceptable, they must come from a sufficiently reliable source, which means that one *could* give good arguments for them. For reasoning to be strong, the premises, if true, must either guarantee or make it sufficiently likely that the conclusion is true. In our argument above, to the effect that soccer is a game, we have acceptable premises and strong reasoning. Not all arguments for such a conclusion, however, are as strong. An argument is not judged on the strength of its conclusion but on the strength of the justification its premises provide. Compare the above argument for (C) in figure 4.2 with the one in figure 4.3.

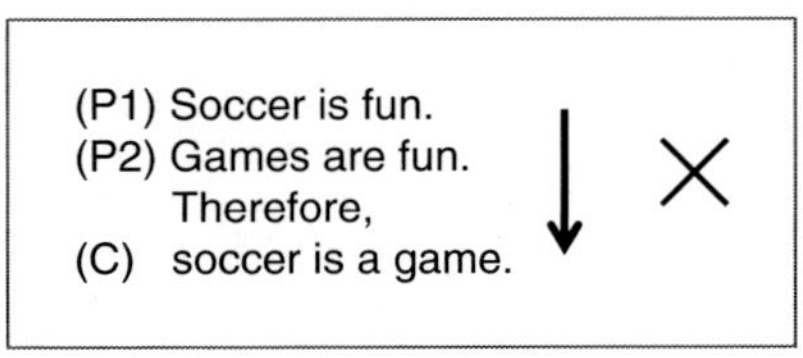

Figure 4.3. Weak Argument
Source: Jason Holt

Our original argument for (C) was cogent, but this one is not. We might find "Soccer is fun" and "Games are fun" to be acceptable premises, but they do not justify (C). Many different things are fun without being the same thing. With respect to our original argument, though, not only should we accept the premises "Soccer is a sport" and "Sports are athletic games," but the reasoning from the premises to the conclusion is airtight in a way I will specify later.

Assessing Arguments

One rather helpful way of analyzing argument **cogency** given by Trudy Govier (2004) involves the mnemonic anagram ARG. Cogent arguments are those that meet all three ARG conditions, with each letter signifying a distinct condition: *A* for **acceptability**, *R* for **relevance**, and *G* for **grounds**. An argument meets the A condition when its premises come from a reliable source. An argument meets the R condition when its premise set bears on the truth of the conclusion. An argument meets the G condition when its premises provide sufficient grounds for the conclusion. Even though the A condition stands alone, conditions R and G are related, in that meeting the R condition is necessary though not sufficient for meeting the G condition. Ensuring relevance is a lower hurdle than providing sufficient grounds. These conditions are nonetheless distinguished in part because many arguments fail to clear that lower hurdle, and so it is often helpful to assess relevance before assessing sufficiency.

Govier's ARG Conditions

An argument has cogency if and only if it meets the following conditions:

- *A*cceptability (of premises)
- *R*elevance (of premises to conclusion)
- *G*rounds (by premises sufficient for conclusion)

Premises are acceptable, again, when they come from sufficiently reliable sources. That is one reason why using reliable sources is crucial when writing essays. According to Govier (2004), a premise meets the A condition if it comes from any of the following sources:

1. True definition (e.g., "Sports require physical skill")
2. Common knowledge (e.g., "Soccer is a popular sport")
3. Trustworthy experience (e.g., normal vision)
4. Genuine expertise (e.g., a professor's understanding of their field)
5. Cogent subargument (e.g., as cited in a peer-reviewed article)

Since a premise's acceptability means a cogent subargument *could* be offered for it, any premise that actually *has* such support is thus also acceptable. What makes an argument a **subargument** is that its conclusion is the premise in a further argument.

Meeting the R and G conditions varies to a certain extent across disciplines, primarily because the appropriate methods of justification vary. Nonetheless, there are certain general patterns of reasoning that are strong regardless of the relevant discipline. Govier (2004) distinguishes four different ways an argument may meet the R and G conditions, four different styles of reasoning to justify conclusions: **deductive entailment** (logic-based reasoning), **inductive support** (statistical reasoning), **analogical support** (likeness-based reasoning), and **conductive support** (cumulative evidence–based reasoning). The last two fall under the rubric of what is sometimes called **abductive reasoning**, which implicitly assume a general explanatory framework and so abduct or "move away from" the particulars. The strength of a piece of statistical reasoning to either a generalization or novel case depends on having a reliable sample, usually one that is sufficiently large and random.[2] Deductive entailment means that *if* the premises are true, the conclusion *has* to be true. Even though inductive, analogical, and conductive arguments never *guarantee* the truth of their conclusions, they can nonetheless provide enough support for the conclusion to meet the R and G conditions:

- *Deductive entailment*
    - Logic-based reasoning
        - (P) Sports require physical prowess.
        - Therefore,
        - (C) figure skating requires physical prowess.
- *Inductive support*
    - Statistical reasoning
        - (P) Most kinesiology students have played sports.
        - Therefore,
        - (C) you, a kinesiology student, have probably played sports.
- *Analogical support*
    - Likeness-based reasoning
        - (P) CTE is a problem in boxing.
        - Therefore,
        - (C) CTE is likely a problem in MMA.
- *Conductive support*
    - Cumulative evidence–based reasoning
        - (P) Jane is a good player, respected by teammates, and a good role model.
        - Therefore,
        - (C) Jane should be team captain.

Forms of Strong Reasoning

- Deductive entailment
- Inductive support
- Analogical support
- Conductive support

One word of caution. Strong reasoning will exhibit one or more of these forms. It does not follow, however, that every inference that seems to fit these profiles will have sufficient strength to meet the G condition. Some reasoning that seems to be logic-based is invalid. Certain statistical or analogical arguments are too weak. You may require practice to learn to discern which is which, especially in the case of fallacies, to which we now turn.

FALLACIES

Justification does not always guarantee truth, but given all available evidence, it is invariably our best indicator of the truth. Successful reasoning provides enough justification for the conclusion given the premises, but this does not happen if the reasoning is erroneous. Such a reasoning error is called a **fallacy**. In fallacies, the pattern of reasoning itself is flawed, and, in Govier's terms, a fallacious argument is one that fails to meet the G condition. Indeed, with many fallacies, the argument fails to meet the R condition as well.

Definition of Fallacy

A fallacy is an error in reasoning.

Fallacies are not only rational failures but personal failures as well. A fallacy may reveal an *intellectual* failure when the arguer fails to appreciate the weakness of their reasoning. Worse, a fallacy may indicate *moral* failure when the arguer is aware of the weakness of their reasoning and is trying to persuade others illicitly. Many fallacies are so common as to be given names as a convenient shorthand. Though I can barely scratch the surface here, I will outline some fallacies kinesiology students should be cognizant of to improve their reasoning and aid in their logical self-defence.

- **Appeal to ignorance**
  - Citing the lack of knowledge or evidence that something is true as evidence that its negation or opposite is true.
    - (P) We have no proof that you are not taking performance-enhancing drugs.
    - Therefore,
    - (C) you probably are taking performance-enhancing drugs.
- **Appeal to popularity**
  - Taking a belief or thing to be true or of value simply because it is common or popular.
    - (P) Conor McGregor is a popular athlete.
    - Therefore,
    - (C) Conor McGregor is a good role model.
- **Post hoc fallacy**
  - Inferring that one event caused another event simply because the former preceded the latter.
    - (P) I stretched before the game in which I was injured.
    - Therefore,
    - (C) stretching predisposed me to injury.
- **Hasty generalization**
  - Taking an inadequate sample, usually one that is too small, as the basis for a more general or universal claim.
    - (P) Joe Genius sailed through biomechanics.
    - Therefore,
    - (C) biomechanics is a bird course.
- **Naturalistic fallacy**
  - This is the fallacy of inferring *ought* from *is*, that is, a value statement from a factual one.
    - (P) Football has been a dangerous sport since its inception.
    - Therefore,
    - (C) the rules of football should not be changed.
- **Normativistic fallacy**[3]
  - An inversion of the naturalistic fallacy: taking a value statement to imply a factual one.
    - **Optimist's fallacy**
      - (P) You deserve to make the team.
      - Therefore,
      - (C) you *will* make the team.
    - **Pessimist's fallacy**
      - (P) You deserve to make the team.
      - Therefore,
      - (C) you will *not* make the team.

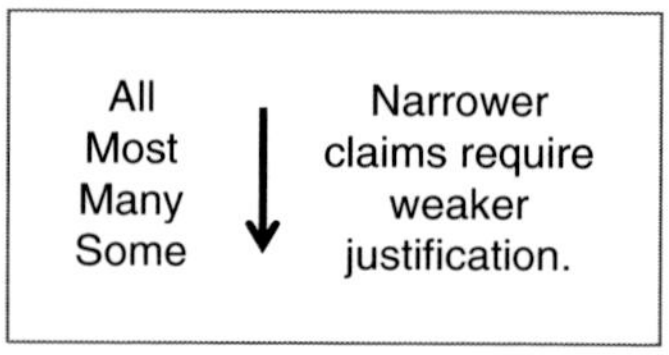

Figure 4.4. Scope Qualifiers
Source: Jason Holt

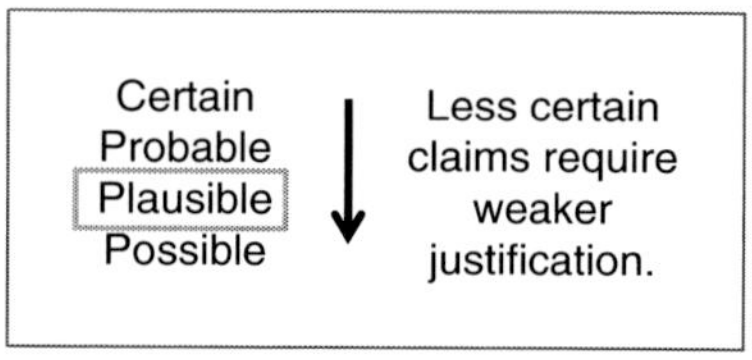

Figure 4.5. Certainty Qualifiers
Source: Jason Holt

There are dozens of other named fallacies, but these, in my experience, are those to which kinesiology students are especially prone. Take care to avoid both committing them yourself and succumbing to those of others.

QUALIFIER FIXES

Suppose your professor assigns you a research-based argumentative essay. Some students in this position will be tempted to overreach, either (1) using premises that are unacceptably strong, (2) drawing conclusions too strong for the premises to justify, or (3) both. You may well be tempted to use the premise "Most athletes overtrain," considering it to be a piece of common knowledge, in arguing for the conclusion "Many conditioning programs are excessive." But it is not common knowledge that *most* athletes overtrain, only that *many* do. Your research likewise will probably justify, at most, a more restricted conclusion, like "*Some* conditioning programs are excessive" or "It is *plausible* that many conditioning programs are excessive." We can transform unacceptable premises into acceptable premises and weak inferences into strong ones by *qualifying* either the scope or the certainty of our claims. Aim for premises weak enough to be acceptable yet strong enough to justify conclusions that are themselves strong enough to be worth arguing for. Weaker claims only require weak justification, but they also provide weaker justification and so less theoretically desirable conclusions. It is a balancing act.

Making a universal generalization about an entire category is theoretically desirable but is also easy to undermine with a **counterexample**: one albino raven falsifies the claim "All ravens are black" but not so the weaker "*Most* ravens are black." Notice in the list of certainty qualifiers that I have highlighted the term **plausible**. This term is absolutely essential as a qualifier in many academic and professional contexts. Whereas *probable* indicates the best bet and *possible* indicates some chance, however slim, *plausible* indicates a good bet: distinct possibility. Where one's evidence is not good enough to make any claim to probability, and mere possibility is less than theoretically desirable, *plausibility* fits the gap beautifully. Right now, I cannot justifiably claim that most university students drink alcohol, though I can claim that this is not only possible but also plausible. Notice that most conspiracy theories, though technically possible, are not plausible enough to be taken seriously.

CRITIQUE AND DEBATE STRATEGIES

Much of the material discussed in this chapter can be used in critically evaluating other people's work and developing debate strategies. In the case of critical evaluation, for instance, you can assess whether an argument is being offered, either explicitly or implicitly. If so, you can use the ARG conditions to assess its quality. Are the premises acceptable? What is the type of inference? Are there evident fallacies at work? Assess the plausibility of the target's position. Can you come up with a counterexample? If so, identify whether the overreach is one of scope or certainty. Might there be relevant evidence left unaddressed?

In debates, you should come up with the best evidence, the strongest arguments you can to support your own side. No less important, however, is being able to counter the other side. In this case, too, look for counterexamples. Think of analogous examples or otherwise undiscussed implications of the other side's position. Ideally, try to incorporate the other side's point of view into your own. Nothing strengthens your own argument like incorporating the opposition's own insights, where possible.

BRIEF RECAP

In this chapter, we covered justification as linking knowledge and its demonstration in argument:

- Knowledge
- Argument
- Fallacies
- Qualifier fixes
- Critique and debate strategies

CHAPTER 5

Writing Essays

CHAPTER OVERVIEW

Essay writing is something you are accustomed to already. If asked, you would probably be at a loss to count how many essays you have written. Still, if you are a first-year university or college student, you might be daunted by how much essay writing lies ahead. However, what is daunting for many need not be overwhelming for any. In this chapter, I hope to leave you with a sense that the task of writing scholarly essays, though challenging, is nonetheless manageable. I plan to do this by reinforcing old advice and giving new advice about how to go about writing scholarly essays. From initial planning to final polishing of your essay, this chapter should help guide and perhaps even motivate you.

We begin with the planning stage, including tips on interpreting essay assignments. Next typically comes the research stage, which includes finding, assessing, and using resources. We end by considering matters of content and structure: the thesis statement, the main body, and finally the introduction and conclusion.

PLANNING

Essay writing begins when your professor gives you an assignment. The assignment will consist of instructions: due date, word count, topic, and so on. The professor will often give you a sense of how marks will be awarded, in the form of either an explicit rubric or more informal guidelines. Too often, students will submit careless work that appears to ignore the given instructions altogether. This is especially puzzling since, obviously, students want good grades, and the easiest part of getting good grades is following instructions. Students who are less than scrupulous in following instructions are just leaving marks on the table.

Either the nature of the course you are writing for or the specific assignment itself should give you a well-defined view of the appropriate topic for your essay, often—though not always—in the form of an essay question. You might be given an array of questions to choose from, and in some cases you may have to come up with the topic yourself. The topic might be very narrowly defined for you, as with "Critically evaluate Suits's theory of games in light of the anti-formalist objection discussed in class." On the other hand, you might be given a general topic: research on the female athlete triad, for instance. Sometimes, too, the field might be completely wide open, as when the professor asks you to come up with your own topic that is relevant to the course, to be approached in a form suitable to your particular subdiscipline. The more open the range of possible topics and the less specifically defined the questions you are asked to answer in essay form, the more it may be prudent, once you come up with an idea for your essay, to discuss the choice with classmates, teaching assistants, or the professor to ensure you are on the right track. A good deal of wasted student time and effort can go into insufficient care of such preliminaries.

Student-chosen essay topics are easier for being less restrictive but harder for the burden of coming up with the topic yourself. Having prescribed topics and questions inverts this double edge. The burden here is less creative than attentive. Scrupulous attention to language is vital in approaching an essay question.[1] Pay close attention to key terms, including verbs, which tell you what tasks you are supposed to perform, as in the following typical examples:

- *outline*
  - state simply, without much development of each point (unless asked).
- *trace*
  - review by looking back—on stages or steps in a process, or on causes of an occurrence.
- *explain*
  - show how or why something happens.
- *discuss*
  - examine or analyze in an orderly way. This instruction allows you considerable freedom, as long as you take into account contrary evidence or ideas.
- *compare*
  - examine differences as well as similarities.
- *evaluate*
  - analyze strengths and weaknesses, providing an overall assessment of worth. (Northey & McKibbin, 2007, pp. 11–12)

Besides verbs, other key terms include technical language. To illustrate, let us consider again the essay question "Critically evaluate Suits's theory of games in light of the anti-formalist objection discussed in class." The verb phrase "critically evaluate" tells you to assess the theoretical power of a particular theory of games, Suits's, limiting your assessment to a particular named objection to that theory—and not just that objection but that objection as framed in a particular way, that is, as covered in class. Your task, therefore, is to argue for a perspective on how damaging a certain objection to a certain theory is to that theory.

To some extent, the latitude you have in selecting topics and approaches will depend on the individual professor as well as their chosen pedagogy, but it will also depend on the subdiscipline of kinesiology you are working in. A research paper in exercise physiology, for instance, will vary significantly from a research paper in the philosophy of sport. In terms of relevant questions, kinds of evidence and argument, and even appropriate language use, standards vary substantially across the different subdisciplines in kinesiology. First-person language is strongly discouraged in the sciences but acceptable and even encouraged in the arts and humanities. So you would avoid the first-person *I* in an exercise physiology essay, but not necessarily in a philosophy of sport essay. In fact, in arts and humanities disciplines that permit the first person, passive constructions can seem stilted:

- Exercise physiology
  - I will critically evaluate the recent literature. ☒
  - The recent literature will be critically evaluated. ☑
- Philosophy of sport
  - I will critically evaluate the recent literature. ☑
  - The recent literature will be critically evaluated. ☒

RESEARCH

When you have a firm grasp on the task at hand and know the instructions for the assignment, the next step is to do research, and doing research often begins—or should begin—with a research plan. The assignment itself will often dictate or set parameters for your research in terms of finding sources, assessing them, and using them. Beyond assignment particulars, this aspect of the research process is also subject to general guidelines.

Although doing research is certainly important, you should also bear in mind that in most cases, even in writing a literature review, you are not just assembling a pastiche of material but using that information to write your own essay. Even when you are asked to critique or review already-established research, that very task adopts a perspective on that literature from the outside but should also be thought of as being both a part of and a contribution to that literature. To write a scholarly essay is not just to report research but to *do* research.

Finding Sources

Before you can assess or use sources, however, you first have to find them. A good place to start is the university library. Searching your library catalogue online is easy enough, but going to the library itself—doing searches there and even walking through the stacks—can be useful in ways that mere online searching often cannot. For instance, online searches may not reveal interesting related material stacked on the shelves according to standard classification systems. If you discover through an online search that a relevant source is only available in hard copy and are forced to make a trip to the stacks, glancing at other books shelved in the same section or nearby may sometimes yield surprising discoveries. Asking for help from on-duty librarians, whether in-person or through the online chat most university libraries offer, can also be helpful.

Your library probably has databases available peculiar to kinesiology and sports studies and to whatever subdiscipline of kinesiology you are working in. Databases like SPORTDiscus are invaluable and specific, but more general databases such as JSTOR can also be useful. Your library will also provide, both digitally and in hard copy, reference resources, including, in some cases, general writing and citation guides and more focused resources that are either discipline- or subdiscipline-specific. Because library search engines are continually being improved, searching for titles and simple strings of words in a general catalogue search will frequently yield a wealth of results, although filtering those results for relevance can sometimes be a challenge.

One way to sidestep the problem of having too many hits come up, whether in a focused library search or a general online search, is to narrow your search by specific journal, particularly—given the subdiscipline in question—top-level or niche journals. In the philosophy of sport, for instance, among various journals that publish articles either in the area or relevant to the area, the two top journals in the world are *Journal of the Philosophy of Sport* and *Sport, Ethics and Philosophy*. A good way to limit preliminary research into the philosophy of sport, whatever your topic is, is to start with those two journals, each of which has its own journal-specific search engine. There are different ways to figure out which are the top-level or niche journals. You could ask a professor, librarian, or students more experienced in the area. You can search relevant terms on Google or Google Scholar to find niche journal titles, if any, and to see which articles in which journals get cited the most. It takes just a little investigation online to find out which articles and journals are having an impact in the relevant subdiscipline.

A related tip is to find, especially in such journals, a recent relevant article on your topic. The bibliography of a recently published article on your topic can give you a head start in finding related relevant sources, since the author(s) of that article should have already done significant background work in the area. This is a helpful shortcut, but don't lean on it as a crutch. Doing so can prevent you from finding even better sources elsewhere, and this approach can therefore be somewhat limiting, especially since many relevant resources will focus on issues different from those of your particular question or topic.

Sometimes it will be prudent to extend your search for sources beyond the scope of those available from or through your library. Depending on your topic, a general online search may be extremely helpful as well. Some of the results obtained from general searching will be unreliable or otherwise problematic, and so this must be done cautiously and with good judgment. Using a search engine like Google can be very productive but can also provide you either with too many results or with results whose reliability is questionable. One way to narrow results of the Google search when typing in a string of words, such as the title of a book or article, is to put the string in quotation marks. That way, the results will include all those words in precisely that order, not just sources containing all of those words in any order whatsoever, usually scattered throughout the document. You can also limit the number of results you get by using not the general Google search engine but specifically Google Scholar. The virtue of this feature is that most of the results it yields are sufficiently scholarly to pass the reliability test.[2] Google Books can also provide access, through the preview feature, to some content of some books that might not be available in your library. Though this is by no means foolproof, it is an extra resource you can use, especially if you know that the book in question is from a reputable, appropriately scholarly publisher.

Finding Scholarly Sources

1. Explore your university library in person, not just online.
2. Search specific websites of relevant top-level and niche journals.
3. Peruse the bibliographies of relevant recent articles and books.
4. Use quotation marks and Scholar to limit Google search results.

Assessing Sources

This naturally raises the question of assessing sources. In some cases, the professor can make this easy for you by limiting acceptable sources cited exclusively to, say, peer-reviewed journal articles. Peer-reviewed journal articles are sometimes touted as the "gold standard" of scholarly reliability. However, you should not dismiss books from scholarly publishers, as these are comparably reliable possible sources. Many essay assignments, though, will require that you refer to sources beyond those that have an obvious academic pedigree. For instance, it may be appropriate to cite articles by journalists in reputable media such as newspapers, most of which are now available in online formats. Although in many cases such sources will prove sufficiently reliable to be worth using as references in a scholarly essay, sometimes it can be hard to distinguish legitimate news from disguised advertisements or even what is sometimes called fake news.[3] Now more than ever, being able to sift sufficiently reliable sources from unreliable ones is of paramount importance.

You have likely been told that crowdsourced resources, such as Wikipedia, should not be used as scholarly references because they are not sufficiently reliable. Nonetheless, it can be hard to resist when a Wikipedia page comes up near the top of most general online searches. To prove to yourself that Wikipedia can be unreliable, explore entries about subjects on which you have some expertise: your hometown, your high school, your favourite athletes, or whatnot. Often enough, you will find that some information given is incomplete, misleading, or outright false.[4] Teachers and professors may tell you to not to use Wikipedia *unless* you can verify its claims with a more reliable source, but I strongly recommend, more stringently, never using it at all—*except* to illustrate the problematic nature of unreliable sources, as a scholarly warning.

Suppose, however, you are faced with an online source that looks really helpful for your essay and yet you cannot tell just by looking at the site whether it is sufficiently reliable. It is not obviously a peer-reviewed scholarly journal or an unreliable crowdsourced article or random blog page. How can you decide if it is reliable? Though there is no foolproof test, the following data will often give you plausible evidence one way or the other:

1. Who is the author and/or the sponsor of the site?
    - Does the site appear under the auspices of a reputable academic or other institution?
    - Is the author (if he or she is identified) someone whose authority you can assess or confirm?
2. Is the site professionally presented?
    - Is the writing well edited and free of errors?[5]
    - Is the site well organized and well designed?
3. Is the site intended for the scholar or for the general reader?
    - Is the text written at a high academic level? Does the author have knowledge of, and make use of, other important sources within the discipline?
    - Is adequate documentation supplied?
    - How does this resource compare with others that are available to you in other formats?
4. Can factual material be verified elsewhere?
    - Are there places, on the Web or elsewhere, where you could go to confirm important facts or statistics?
    - What is the nature of the links provided? Are they evaluated?

5. What does the URL (the internet address) tell you about the site?
    - A .com (commercial) site exists principally to generate income for a company.[6]
    - A .edu site originates at a US university.
    - A .org site belongs to a (usually non-profit) group.
    - Other designations include .net (network), .gov (government), .mil (US military), and .ca (originating in Canada); there are similar short forms for other countries.
6. Is the site current or recently revised?
    - Is someone keeping the site current, or has it been installed and then abandoned? (Heckman, 2007, p. 30)

If, after examining the source with these questions in mind, you remain unable to tell whether the source is sufficiently reliable, you cannot just assume that it *is*, so you should assume that it is *not*. This advice may appear inconsistent with the principle of charity: that in interpretating texts you should give authors the benefit of the doubt. But an ethical presumption of *honesty* does not imply a corresponding presumption of *reliability*.

Using Sources

Reliability is only part of the picture, however. Reliable sources may not be particularly relevant either to the topic you are writing on or, more narrowly, to the perspective you are arguing for. It is not always easy, either, to tell whether a potential source is relevant or, if so, which information is worth using for your essay. In this way, relevance factors into both potential source assessment and information selection within relevant sources.

But what is relevance? Relevance is a relation between two pieces of information or data such that one *bears on* the other. It comes in two forms, which we can call "topic relevance" and "truth relevance," respectively. Take x and y as variables: x is *topic*-relevant to y just in case x bears on the subject of y; x is *truth*-relevant to y just in case x bears on the justification of y. This is a bit abstract, but some examples will help. Suppose the subject of discussion is the theory of relativity:

- An example of **topic relevance**:
    - The theory of relativity *was developed by* Albert Einstein.
- An example of **truth relevance**:
    - The theory of relativity *has been confirmed by* empirical evidence.

That Einstein developed it bears on the theory of relativity as a subject of discussion, though not on whether we have cause to believe that theory. *Topic relevance fails to imply truth relevance.* This is an important point and distinction because students often fail to appreciate the difference. Often students will err by padding their essays with far too much information that bears only on the topic—and not also on the truth—of the viewpoint they are trying to establish. The result, in most cases, is an ill-structured and weakly argued essay. By the same token, material that counts *against* the perspective you hope to establish is no less truth-relevant than supporting material—and in some sense is even more significant. Any essay will be stronger if it parries arguments to the contrary.

You should use sources responsibly, and at the heart of responsible scholarship is acknowledging your sources. The basic concept of plagiarism is familiar already: presenting the work of someone else as if it is your own. More precisely, I will define **plagiarism** as *presenting material as if it were one's own original work when in fact it is not.* Plagiarism includes various types of mistakes and is a violation of academic integrity irrespective of the offender's intentions. That is, whether you mean to commit plagiarism or not fails to bear on whether, in fact, you have committed plagiarism. Here are different varieties of plagiarism:

- Copying text (including passages, sentences, or phrases) from someone else without both quoting and citing it.
- Changing a few words in a quotation and presenting it without a citation or as if it were a paraphrase (i.e., written in your own words).
- Using specific information (besides common knowledge) from a source without citing it.
- Presenting someone else's ideas as your own (whether specific examples and arguments or general theories and approaches).
- Submitting an essay you paid someone else to write (including the products of custom essay writing services) as your own.
- Submitting the same—or a sufficiently similar—essay written for one course for credit in another course (i.e., **self-plagiarism**).[7]

Definition of Plagiarism

Plagiarism is presenting material as if it were one's own original work when, in fact, it is not.

Students relying on only a basic understanding of plagiarism can find the idea of self-plagiarism paradoxical, since you are not presenting someone *else's* work as your own. In the same way, an essay writing service may "guarantee" essays that are "100 percent plagiarism-free." Neither loophole is genuine, however. Work you did for one course is no longer *original* work for another course, and though the anonymous author of a custom essay may not have committed plagiarism, when you submit it as your own, *you* have. In general, using words, information, or ideas that are already on record as if you are generating them originally counts as plagiarism. That is the key reason why proper documentation is so important.[8]

The Three *R*s: Assessing and Using Sources

- *R*eliability
  - Sources used must be sufficiently reliable both in general and as required by the given subdiscipline.
- *R*elevance
  - Both sources and specific source material used must be not just *topic*-relevant but also *truth*-relevant to the perspective being supported.
- *R*esponsibility
  - All source material (including text, information, and ideas) must be acknowledged properly using an acceptable documentation style.

When introducing support material, pay attention to your writing mechanics, especially punctuation and usage. Say you want to introduce a quotation that identifies the conclusion that your source has argued for. Use a comma when your introducing language is not an independent clause, a colon when it is an independent clause, and no punctuation when the quote's grammar fits the grammar of the sentence:

- As Holt argues, "some figure skating performances count as both sport and art" (2017, p. 145).
- Holt argues for a restricted position: "some figure skating performances count as both sport and art" (2017, p. 145).
- Holt argues that "some figure skating performances count as both sport and art" (2017, p. 145).

Likewise, use introducing language that conveys something of the appropriate attitude, your own or that of the author(s) of the source. Often you should prefer a neutral description, using *states*, *argues*, *asserts*, *claims*, and so on. Sometimes it may be better, however, to imply something about your or the source author's attitude. Some examples follow:

- *Neutrality*: "Holt *states* that ..."

- *Consensus*: "Holt *agrees* that ..."
- *Endorsement*: "Holt *shows* that ..."
- *Skepticism*: "Holt *speculates* that ..."
- *Commitment*: "Holt *insists* that ..."

When introducing support material, then, make sure not only that punctuation matches grammar but also that your language fits both what you want to say and what you want to further suggest. All else being equal, whenever a word choice conveys more intended meaning, it is the better option.

CONTENT AND STRUCTURE

As for any text you compose, your essay has a beginning, middle, and end. The beginning is the introduction, the middle is the main body, and the end is the conclusion. Before you start in on these, however, your first concern should be the heart.

Thesis Statement

The heart of your essay is the **thesis statement**, which should appear in your introduction as well as your conclusion. Your thesis is the point of view you are trying—essaying—to establish in the essay. As a statement, a thesis is expressed as an assertion, the sort of sentence that can be true or false and that you try to justify in the course of the essay. Sometimes you might have a good idea of what your thesis will be before you even do research—say, that performance-enhancing drugs should remain prohibited in sports. In other cases, though, it may take a considerable amount of research and development of ideas before you even begin to formulate a thesis statement: if your topic is optimization models in golf, you might not know until you get deep into the research what the topic's parameters are, much less the stance you will take. Reviewing research materials will give clues about what research questions are worth asking and what kind of thesis statement will be relevant and manageable to establish in your essay.

Definition of Thesis Statement

A thesis statement is the unifying summary statement of an essay intended to be supported by the main body.

At whatever stage of the process you formulate your thesis statement, a good one will be marked by certain features. First, your thesis should be relevant to the subdiscipline and the specific essay assignment. Because kinesiology is interdisciplinary, students may sometimes neglect the fact that a thesis statement relevant in one subdiscipline will often prove irrelevant in another. Relevance aside, your thesis statement should also be appropriately restricted, unified, and precise (Northey & McKibbin, 2007). A restricted thesis is one that you can manageably support within the confines of the assignment. In a short essay, for instance, you probably will be unable to adequately support a thesis covering all the available research on, say, the role of nutrition in performance enhancement. A unified thesis statement makes a single key point rather than two or more. If your research yields some evidence in favour of stretching as a means of preventing injuries, and some evidence against this, your thesis can be unified either in assessing which evidence is stronger or concluding that it is inconclusive. If you have a handful of main points to make, express your thesis in a general form covering them all. As for precision, your thesis statement should suggest to the reader what you mean and what sort of evidence would be needed to justify the statement. Instead of a vague thesis statement about, say, conditioning being good for performance, be as precise as you can about the specific type of conditioning you mean and the specific type of performance your evidence supports. In general, your thesis statement is best phrased as an answer to a significant question, that question being either explicitly asked by the assignment—the essay question, if given—or implicitly asked by you.

Thesis Statement Essentials

A good thesis statement is

1. Relevant (to the subdiscipline and specific assignment)
2. Restricted (so it can be supported manageably within assignment parameters)
3. Unified (with subsidiary points prioritized or subsumed)
4. Precise (in meaning, with clear evidentiary implications)
5. Responsive (to an explicit or implicit essay question)

Suppose you are writing an essay in the philosophy of sport on the question of whether sport can be art. You may think to yourself that there is a connection between works of art and enjoyable experiences, but that alone fails to yield a good thesis statement:

- *Poor thesis statement (fails criteria 1–5)*
    - Many people find sport enjoyable in countless ways, but many others do not.

- *Good thesis statement (meets criteria 1–5)*
    - Against Best's "life situations" argument that sport cannot be art, I draw on the work of Peter Arnold and Jason Holt to argue that, in the case of figure skating, sport can be art.

In the case of the poor thesis statement, it is not clear that the question is relevant to philosophy of sport as opposed to, say, psychology of sport. Also, it is neither restricted nor unified, nor is its meaning precise or obviously responsive to the question of whether sport can be art. On the other hand, the good thesis statement is clearly relevant to the philosophy of sport, manageably restricted, unified, and responsive to the sport-as-art question.

Main Body

With most of your research done, before you write your first draft you should construct a blueprint of the essay to come: an outline. But to do this, you will usually need to organize your source material and develop your ideas first. If you do not yet have a set thesis statement, this preliminary work can help you formulate and refine one. More importantly, such preparation is the key precursor to a well-structured and successful first draft.

Having finished note-taking from sources does not necessarily mean that you are ready to start outlining. You need to develop the ideas at your disposal, those of the source material along with your own. This may involve, for instance, defining terms, categorizing items, making sense of processes, finding causes or effects, and drawing comparisons (Northey & McKibbin, 2007, pp. 23–26). You can also develop ideas by finding or abstracting from examples and exploring logical relations. Think of analogies and disanalogies between domains (e.g., how kick-boxing is and is not like MMA), understanding things in terms of their intended purposes (e.g., the motive behind sport spectacles), and critical and creative thinking (e.g., questioning assumptions about the value of sport and whether it is worth pursuing). Not all of these approaches will be useful in all contexts, but they are options to keep in mind, especially if you get stuck.

When defining key terms—which is always prudent in formal essays—you should avoid general dictionary definitions in favour of (sub)disciplinary definitions. The best general lexicon of the English language, the *Oxford English Dictionary*, though great in most contexts, is simply not precise enough to provide apt definitions of technical kinesiological terms. For example, the *OED*—its affectionate acronym—defines *flexibility* as "capability of being bent; pliancy" and "susceptibility of modification or alteration; capacity for ready adaptation to various purposes or conditions; freedom from stiffness or rigidity" (Oxford University Press, 2021). This is nowhere close to disciplinary definitions, such as "the intrinsic property of body tissues that determines the range of motion achievable without injury at a joint or group of joints" (Holt et al., 1996, p. 172). Look for (sub)disciplinary definitions in kinesiology textbooks and journal articles that focus on your specific topic.

Ways of Developing Ideas

1. Define terms.
2. Categorize items.
3. Explain processes.
4. Look for causes or effects.
5. Draw comparisons.
6. Look for or abstract from examples.
7. Explore logical relations.
8. Draw analogies and disanalogies.
9. Look for underlying purposes.
10. Use critical and creative analysis.

Outlining before you start to write is essential in most forms of formal writing, especially essays, where it is of paramount importance. An outline is often useful, but for a formal essay it is essential. A good first step in outlining is to organize your research material, including your own ideas, into chunks of information. This "chunking" procedure will suggest how to group the material into different sections or paragraphs. Sometimes a topic divides neatly into subtopics. If you are writing about stretching techniques, for instance, your source material might chunk into the three main varieties of technique: slow/static, dynamic, and reversal of antagonists. In certain cases, though, chunking depends on you imposing a logical order on the material. One chunk might concern studies that have found a certain result, another studies to the contrary, and a third an analysis of the debate. When you have a sense of potential chunks, this is a good basis for finding a unifying theme or thesis statement—and more important for assessing and prioritizing the evidence you have for your thesis. This procedure can also help you identify gaps, order chunks logically, and refine your thesis statement. Be receptive to discarding chunks that fail to cohere your emerging theme and thesis. The chunks retained should either suggest or lend themselves to logical ordering. For instance, if you are writing about a process, it will often make sense to structure chunks to parallel the process itself (background conditions, then causes, then immediate effects, and finally long-term effects). If you are critically evaluating a theory, it will usually make sense to follow the theorist's original order of presentation. Where each chunk is a different piece of evidence or argument for your thesis statement, you may order by strength (weakest to strongest), or novelty (least to most original), among other approaches.[9]

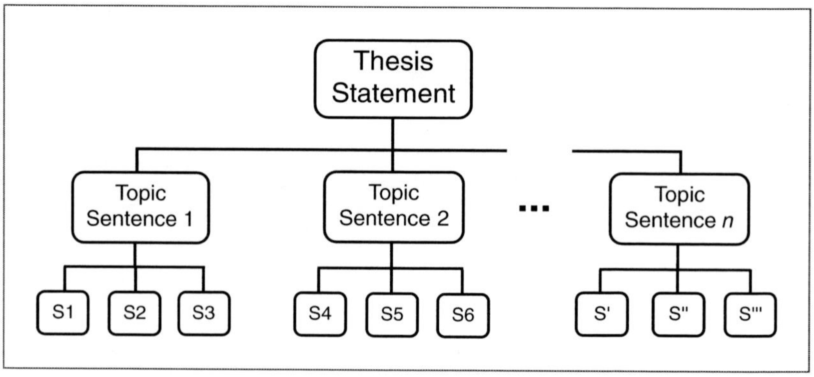

Figure 5.1. Structural Parallel (Essays)
Source: Jason Holt

Once you have your chunks ordered, you are effectively done with the first layer of your outline: a thesis statement and supporting chunks. At this stage, try to come up with sentences to encapsulate the main point of each chunk. Each of these statements will serve a dual role, both as principal support for your thesis statement *and* as the **topic sentence** for a paragraph—or what could be called the "sub-thesis statement" of a section that itself needs unpacking into separate paragraphs. Each topic sentence can then be used to unify and order the other sentences in each paragraph on a fine-grained level the way, at a coarse-grained level, the thesis statement should govern the topic sentences of supporting paragraphs and, in effect, the whole essay. Take seriously, then, the following analogy: *the thesis statement is to the essay as the topic sentence is to the paragraph.* The thesis should unify the topic sentences just as the topic sentences unify the individual paragraphs. With this overall structure in place, your outline is complete.[10]

Introduction and Conclusion

The introduction and conclusion bookend the **main body** of your essay. Some essayists prefer to write the introduction and sometimes even the conclusion first, before drafting the main body. In either case, I will assume that you have done all or most of your research and have a preliminary, if not full, outline in hand as well as a thesis statement.

One technique that you may find useful is to write your introduction as if it were, more or less, an **abstract**: a paragraph summary of the entire essay, including not just the thesis statement but the main lines of support for it. Indeed, some professors find that writing an abstract is a key step in any research project, even if the abstract is ultimately not part of the published work. For many of my research projects, for instance, I begin by answering a conference call for abstracts, submitting an abstract for a planned but not yet written work. If the abstract is accepted, I create from the abstract a slideshow presentation that gives essentially a detailed outline of the essay to come. After presenting at the conference and incorporating any feedback, I write the essay itself before submitting for publication.[11] You may or may not have such a long view in mind, but it could turn out that writing an abstract is of comparable help to you.

Your introduction raises a topic of discussion, puts it in context, outlines how you plan to approach it, and expresses, in the form of a thesis statement, the perspective you aim to support. You should open, if you can, with a hook, something that grabs the reader's attention. You might, for instance, use a quotation, a rhetorical question, an anecdote, or a fact worth noting (Northey & McKibbin, 2007, pp. 27–28). The quotation might be the very first line of your text, or it could appear before the text itself, as an **epigraph**. In using rhetorical questions, try to avoid clichés unless they are unusually pertinent to your topic. An anecdote can be drawn from real life, from your own or someone else's actual experience, or can be hypothetical and flagged as such, using "Suppose ..." and the like. Noteworthy facts may be surprising or revealing and should concern either statistical trends or specific known cases. After your opener, think of your introduction as funneling from a broad topic to a specific thesis (Baker & Gamache, 1998, p. 65, as cited in Northey & McKibbin, 2007, p. 26). The context is the kinesiological subdiscipline in which you are working and the questions it deems relevant, whereas the approach depends on how you frame the question and the research methods used. How you write these up will depend on the subdisciplinary conventions in effect.

Five Elements of Essay Introductions

1. Opener
2. Topic
3. Context
4. Approach
5. Thesis

You may have heard the cliché that, in an essay, first say what you will say (introduction), then say it (main body), and then say what you said (conclusion). This sounds silly, but there is a point to this. Your conclusion is not simply the end of the essay, not simply a summing up—though it is these as well—it should also be understood as serving to reinforce your thesis statement and the justification you provided for it. Where an introduction narrows a broad topic through context to a specific thesis statement, think of the conclusion as inverting this, as beginning with a restatement of the thesis, then expanding to outline, briefly, its implications and broader significance.

Ideally this will include open questions raised by your thesis as taken for granted along with suggestions for further research. End with a decisive, memorable closer. You can refer back to your opening hook, use a stylistic or poetic flourish, or perhaps do something else altogether. See what works for you. Experiment.

Five Elements of Essay Conclusions

1. Thesis
2. Broader significance
3. Open questions
4. Further study
5. Closer

Further Tips on the Essay-Writing Process

1. Follow instructions.
2. Choose a topic based on your own emotional response (positive or negative).
3. Give yourself time.
    - Aim to start and finish early.
    - Break the project into small, manageable tasks.
    - Take a break before revising.
4. Focus on truth-relevant source material.
5. Demonstrate your knowledge to the reader.
6. Proofread carefully during and after editing.
7. Be tech-paranoid (back up work in progress).

BRIEF RECAP

In this chapter, we covered different elements of essay writing:

- Planning
- Research
- Content and structure

CHAPTER 6

Documenting Sources

CHAPTER OVERVIEW

In scholarly work it is crucial to document your sources, not only to avoid violating principles of academic integrity, but also to put your own work in proper context. Your scholarly work should aim to do more than just assemble other people's research. Think of adding something, however modest, to the conversation. Documentation helps the reader follow the conversation.

After a brief introduction of different types of documentation styles, we discuss different in-text citations, specifically quotations and paraphrases. We flag what, in my experience, are the five most common mistakes students make with in-text citations, and then we move on to bibliography basics. We follow this with a list of the 10 most common errors students make in reference lists. We then end by briefly comparing how specific references are formatted in APA in comparison with other documentation styles.

DOCUMENTATION STYLES

Documentation styles can nearly always be divided into two distinct parts: (1) specific references in the text itself and (2) a bibliography at the end. Depending on the style, specific references come in two varieties: **in-text references** or **noted references**. In-text references are **author-date** or **author-page**, and noted references appear in either footnotes at the bottom (foot) of the page or endnotes at the end of the text but before the bibliography. Typical bibliographies are ordered alphabetically by author surname, but some styles require non-alphabetical sequencing (e.g., by order of in-text reference). In certain cases, there is no separate bibliography. Instead, the first noted reference to a given work will contain complete bibliographic data, with subsequent references appearing in a standard abbreviated form.

Documentation Styles

1. Specific references
    - In-text
    - Noted
        - Footnotes
        - Endnotes
2. Bibliography
    - Alphabetical
    - Non-alphabetical (e.g., in order of in-text reference)

Of the many different documentation conventions available, one of the most common in the various subdisciplines of kinesiology is APA (American Psychological Association, 2020), an author-date style. In APA style, specific references are given in-text, usually at least partly in parentheses, and the bibliography is called "References." Below are some of the basics you need to know about APA citation.

QUOTATION VERSUS PARAPHRASE

For specific references, you may choose to either quote or paraphrase the cited work. Compare the following typical examples:

- *Quotation*
    - Holt (2017) argues that "some figure skating performances count as both sport and art" (p. 145).
- *Paraphrase*
    - Holt (2017) argues that in figure skating, sport can coincide with art.

In both cases, the date of publication of the cited work is given in parentheses, usually after the author's surname. This is typical of author-date styles. When you quote rather than paraphrase, the quoted passage is enclosed in quotation marks, and the page number is also given in parentheses.

Students often prefer to paraphrase, in some cases perhaps because it tends to give the impression that they have done more of the work. However, it is sometimes better to quote the original passage, especially if it is well-written, controversial, or otherwise noteworthy. Students also often fall prey—due to inattention or temptation—to the unfortunate shortcut of inadequate paraphrasing: changing the odd word or phrase and keeping the original structure and then offering the result as if written in their own words. Here are some examples:

- *Original quotation*
  - Holt (2017) argues that "some figure skating performances count as both sport and art" (p. 145).
- *Attempted paraphrases*
  - Holt (2017) argues that some figure skating performances are both sport and art. ✗
  - Holt (2017) argues that some figure skating counts as art and sport. ✗
  - Holt (2017) argues that some on-ice activities count as both sport and art. ✗
  - Holt (2017) argues that in figure skating, sport can coincide with art. ✓
  - Holt (2017) argues that in figure skating, sport and art may be found together. ✓
  - Holt (2017) argues that sport and art can sometimes overlap in figure skating performances. ✓

Merely omitting or changing a few words will not result in adequate paraphrasing. You need to change the sentence structure as well as the wording itself. This is *not* a mechanical task but one requiring effort and skill to grasp an idea and express it differently. Good paraphrasing implies an interpretive and creative facility with language.

Adequate Paraphrasing

Adequate paraphrasing requires rewriting, not just changing a few words, even of short passages.

IN-TEXT VARIATIONS

Not all APA in-text citations will fit into the simple format above. Here are some of the common variations on the basic format outlined above:

- If your sentence omits the author's name, that name should appear in the citation itself: "(Holt, 2017, p. 145)" for the quotation, "(Holt, 2017)" for the paraphrase.
- If a quotation continues from one page to another, indicate the relevant page range: e.g., "pp. 145–146."[1]

- For electronic sources without pagination, use "para." to indicate paragraph number in place of the "p." for page, and likewise use "paras." for ranges.[2]
- For longer works, the paragraph count should fall within specified divisions such as article sections, as in "(Holt, 2017, 'The artform question,' para. 3)" rather than "(Holt, 2017, para. 23)."
- Use "and" for two-authored works when the names appear in a sentence, as in "Holt and Holt (2010)," but use "&" otherwise, as in "(Holt & Holt, 2010)."
- When a work has three or more authors, mention the first author's name followed by "et al." (short for *et alia*, from the Latin for "and others"), as in "Holt et al. (2010)" when "Holt et al." is part of the sentence, and "(Holt et al., 2010)" otherwise.
- When citing two or more works that make the same point, use a semicolon, as in "(Holt, 2017; Parry, 2019)."
- When citing multiple works by the same author, use a comma between publication years, as in "(Holt, 2017, 2019)."
- If you cite multiple works from the same author that were published in the same year, use lowercase letters to disambiguate them according to bibliographical order (alphabetized by title), as in "(Holt, 2019a, 2019b)."

Page or Paragraph?

In APA style, for in-text quotations

- Use "p." or "pp." as appropriate for page number(s) when hard-copy pagination is available.
- Use "para." or "paras." as appropriate for paragraph number(s) when pagination is unavailable.

For longer quotations, remember the **40+ Rule**: quotations of fewer than 40 words should be indicated with quotation marks, and quotations of 40 words or more must be set off as **block quotations** (*not* in quotation marks!).

- *Short quotation* (less than 40 words)
  - Holt (2017) argues that "some figure skating performances count as both sport and art" (p. 145).
- *Block quotation* (40 words or more)
  - Holt (2017) draws the following conclusions:

    (1) some figure skating performances count as both sport and art; (2) some dance performances count as both art and sport; (3) figure

> skating (if no other sport) counts as an artform; (4) competitive dance (if no other sort) counts as a sportform. (p. 145)[3]

The significance of these conclusions …

The 40+ Rule

In APA style quotations of 40 words or more must be set off as block quotations.

When introducing any quotation, short or block, use only such punctuation as would be required by the grammar of the introducing language, treating the quotation as part of the sentence (hence no punctuation after *that* and a colon after *conclusions* in the examples above, a function not of the quotation type but rather of the introducing language). Following a block quotation, avoid indenting the text unless it makes sense to start a new paragraph. Usually it will not make sense, especially if you then comment on the passage, which you usually should.

TOP FIVE STUDENT ERRORS (IN-TEXT CITATIONS)

Some errors in APA in-text referencing stand out as particularly common in student scholarship and are thus worth mentioning so you can avoid them. Obviously the most significant blunder is *failing to provide a citation when one is required*. This is plagiarism, and plagiarism is an ethical problem, a moral failing of the student. Because this is a failure to cite *at all* rather than a failure to cite *correctly*, the following list assumes that wherever an in-text reference is needed, students have attempted to provide one. These errors may seem unimportant, but if you avoid them, you can give an impression of professionalism rather than sloppiness. If for no other reason than the sake of making a good impression, note and avoid these errors:

1. *Quotation without page (or paragraph) number*
   - Holt (2017) argues that "some figure skating performances count as both sport and art." ✗
   - Holt (2017) argues that "some figure skating performances count as both sport and art" (<u>p. 145</u>). ✓
2. *Nonspecific page or range*
   - Holt (2017) argues that "some figure skating performances count as both sport and art" (<u>pp. 138–145</u>). ✗
   - Holt (2017) argues that "some figure skating performances count as both sport and art" (p. 145). ✓

3. *Confusion of "p." and "pp."*
   - Holt (2017) argues that "some figure skating performances count as both sport and art" (pp. 145). ☒
   - Holt (2017) argues that "some figure skating performances count as both sport and art" (p. 145). ☑
4. *Reference after punctuation*
   - Holt (2017) argues that "some figure skating performances count as both sport and art". (p. 145) ☒
   - Holt (2017) argues that "some figure skating performances count as both sport and art" (p. 145). ☑
5. *No space before page number*
   - Holt (2017) argues that "some figure skating performances count as both sport and art" (p.145). ☒
   - Holt (2017) argues that "some figure skating performances count as both sport and art" (p. 145). ☑

Avoiding these citation errors is no guarantee that you will avoid errors altogether. Making sure you avoid them, however, is the first leg on the journey to proficiency.

BIBLIOGRAPHY BASICS

At the end of your work comes the bibliography, a list of references used. These references give complete citation data for your sources. What "complete" means in this context depends on which documentation style is in effect. Only some styles, for instance, require references to include the place of publication for books. All styles require bibliographic references to include, if available, the name of the author(s), year of publication or production, title and other features that identify the work, and where, in some sense, the work may be found. Often for a shorter work contained within a larger one—whether an article in a periodical, chapter in a book, online clip from a movie or TV show, song from an album, etc.—both titles are mentioned along with identifying features of both works. Self-contained works such as books, movies, and stand-alone websites are governed by their own conventions.

For APA style, the bibliography should begin on a new page, with the title "References" set in bold and centred. Like the text itself, the references should be double-spaced but without additional line spaces between references. References are visually distinguished by *hanging* indents, which is the inverse of normal paragraph indenting, leaving the first line unindented and then indenting each subsequent line of the same individual reference. Most word processors have a "hanging indent" option that you can use to format your reference list. Entries should be listed alphabetically by author surname. Use initials rather than authors' full names even if the given name appears in the source. If the author is given as "Jason Holt," the bibliographic entry will begin "Holt, J." If you cite multiple works by the same author, list them chronologically. If no author is given (not even an organization), the title takes the place of the author's name. If no date is given, use "(n.d.)" for "no date" instead of the year in parentheses.

Typical bibliographic entries in APA style will have the following elements, in this order: author name, year, specific work title, larger work title (where the specific work is located), identifying elements of the larger work (e.g., journal volume and issue number), identifying elements of the shorter work (e.g., article page range). Specific work titles are set in normal (called "roman") type, whereas larger work titles are given in italics. Capital letters are only used in titles of books, articles, and chapters for the first letter, the first letter following a colon, and the first letters of names. When citing works such as a journal article, indicate the particular volume and issue of the journal where the specific article is found as well as the page range for that article. Where a **DOI** (digital object identifier) is provided for the source, include it at the end. It is important, however, when you access a work online, such as a PDF version of an article, not to omit the available print copy citation data or just give a URL. Giving a URL (or DOI) without hard-copy citation data is only appropriate for exclusively web-based sources.

Here are some examples:

- *Article (print data)*

  Holt, J. (2017). Sport as art, dance as sport. *AUC Kinanthropologica, 53*(2), 138–145. https://doi.org/10.14712/23366052.2017.11
- *Article (online only)*

  Holt, J., & Wills, B. (2019, November 8). Thresholds, defaults, and Cosbys: A response to Bartel. *Contemporary Aesthetics, 17.* https://www.contempaesthetics.org/newvolume/pages/article.php?articleID=880
- *Chapter in edited book*

  Holt, J., & Holt, L. E. (2010). The "ideal" swing, the "ideal" body: Myths of optimization. In A. Wible (Ed.), *Golf and philosophy: Lessons from the links* (pp. 209–220). University Press of Kentucky.
- *Book*

  Holt, J. (2019). *Kinetic beauty: The philosophical aesthetics of sport.* Routledge.
- *Video (online)*

  Holt, L. E., & Holt, J. (2019, July 2). *The cult of conditioning and the harm it does* [Video]. YouTube. https://www.youtube.com/watch?v=V-UYMAdx3y0

When in doubt, you need only a little effort to look up the correct formatting for a given kind of bibliographic entry. Your university library's website will be helpful, as will perusing the relevant style manuals, in this case APA, in your library's reference section. If you cannot find the exact type of case you need to cite, my advice is to proceed by analogy with known cases. You might cite, for example, a song on an album or a poem in a book by analogy with the above convention for book chapters.

TOP 10 STUDENT ERRORS (REFERENCES)

Not all professors will be equally thorough in evaluating how strictly you follow conventions for APA bibliographies, especially after wading through the body of your essay or lab report among who knows how many others. Like officials in competitive sports, professors enforce such rules more or less restrictively for any number of reasons. Just as you would, or should, respect the call of an official in sport, you should respect your professors in enforcing citation rules. Of course, like officials, professors are imperfect, and that is precisely the point. Although some markers will be less concerned than others with certain of the following errors, you should be aware of them nonetheless.

1. *Improper heading*
    - Labelled "Bibliography" or "Works Cited" instead of "References"
2. *Improper order*
    - Not alphabetized by author surname but randomly or otherwise ordered
3. *Improper spacing*
    - Single- rather than double-spaced and/or spaces between entries[4]
4. *Missing data*[5]

    Holt, J. (2017). Sport as art, dance as sport. *AUC Kinanthropologica*, 138–145. ✗

    Holt, J. (2017). Sport as art, dance as sport. *AUC Kinanthropologica*, *53*(2), 138–145. ✓
5. *Misused capitals or lowercase in titles*

    Holt, J. (2017). Sport as Art, Dance as Sport. *Auc kinanthropologica*, *53*(2), 138–145. ✗

    Holt, J. (2017). Sport as art, dance as sport. *AUC Kinanthropologica*, *53*(2), 138–145. ✓
6. *Incomplete page range*

    Holt, J. (2017). Sport as art, dance as sport. *AUC Kinanthropologica*, *53*(2), 145. ✗

    Holt, J. (2017). Sport as art, dance as sport. *AUC Kinanthropologica*, *53*(2), 138–145. ✓

7. *Misused italics for shorter works or roman type for longer works*

   Holt, J. (2017). *Sport as art, dance as sport.*
   AUC Kinanthropologica, *53*(2), 138–145. ✗

   Holt, J. (2017). Sport as art, dance as sport.
   AUC Kinanthropologica, *53*(2), 138–145. ✓

8. *Misused roman type for volume or italics for issue*

   Holt, J. (2017). Sport as art, dance as sport.
   AUC Kinanthropologica, 53*(2)*, 138–145. ✗

   Holt, J. (2017). Sport as art, dance as sport.
   AUC Kinanthropologica, *53*(2), 138–145. ✓

9. *Space between volume and number*

   Holt, J. (2017). Sport as art, dance as sport.
   AUC Kinanthropologica, *53* (2), 138–145. ✗

   Holt, J. (2017). Sport as art, dance as sport.
   AUC Kinanthropologica, *53*(2), 138–145. ✓

10. *No hanging indent*

   Holt, J. (2017). Sport as art, dance as sport.
   AUC Kinanthropologica, *53*(2), 138–145. ✗

   Holt, J. (2017). Sport as art, dance as sport.
   AUC Kinanthropologica, *53*(2), 138–145. ✓

Pro Tip: Citation Generators

Avoid or double-check automatic citation generators, which often omit essential citation data.

FOR COMPARISON

I have focused on APA style, since it predominates in many subdisciplines of kinesiology. Still, looking briefly at other styles will be worthwhile, not only for comparison, but because students will encounter other styles outside of kinesiology (if not also within it). Furthermore, abstracting from the features of any single style will help you better grasp variations among and elements of these important if somewhat arbitrary conventions.

Here are some different styles for specific paraphrase citations. (The original quotation is from Holt's 2017 article "Sport as Art, Dance as Sport": "some figure skating performances count as both sport and art" [p. 145], as above.)

- *APA*
  - It has been argued that in figure skating, sport can coincide with art (Holt, 2017).

- *MLA*
  - It has been argued that in figure skating, sport can coincide with art (Holt 145).
- *Chicago*
  - It has been argued that in figure skating, sport can coincide with art.[2] [*Note*: 2. Holt, "Sport as Art," 145.][6]
- *CSE*
  - It has been argued that in figure skating sport can coincide with art.[1] [*References*: 1. Holt J. 2017. Sport as art, dance as sport. AUC Kin. 53(2):138–45.][7]

Note that where APA stresses the date, MLA and Chicago styles stress the page. Note, too, how a noted style like Chicago and CSE keeps the text relatively clean by minimizing in-text indication of specific references. Having these different conventions at the back of your mind can help you manage most of the minutiae of APA style.

BRIEF RECAP

In this chapter, we covered the essential points of scholarly documentation of sources using styles such as APA:

- Documentation styles
- Quotation versus paraphrase
- In-text variations
- Top five student errors (in-text citations)
- Bibliography basics
- Top 10 student errors (references)
- For comparison (APA and other styles)

CHAPTER 7

Usage Issues

CHAPTER OVERVIEW

In this chapter, we cover various issues concerned with usage, including informally common and successful use of terms that in formal contexts are incorrect or simply poor form. We begin with a number of common nonwords that should be avoided in favour of the corresponding words for which they can be confused. We then look at my list of the top 20 usage confusions, followed by tricky singular/plural pairs in academia broadly and kinesiology in particular. Next we address another dozen common confusions, ending the chapter with tricky word/phrase pairs.

INFORMAL VERSUS FORMAL USAGE

In everyday contexts, nothing will go terribly wrong if you write "In every day contexts …" rather than "In everyday contexts …" as I just did. Many people will miss the mistake. (Did you catch it right away, or at least notice the difference?) Those who catch it might not care. Even those who do care will still know what you mean. However, in formal writing and speaking, both academic and professional, usage mistakes count—a lot. They give the impression that you are careless, if not ignorant of good form and proper usage, which makes people—rightly or wrongly—less receptive to your message. The result is lower grades for students and diminished credibility for professionals.[1] No one wants that. Others may be wrong to judge us poorly because of poor language use, but they often will nonetheless, and it behooves us as communicators to reduce the opportunity for others to undervalue what we say. If others fail to appreciate our *intended* meaning, that is often because what words mean is determined *not* by our intentions but rather by rules and conventions we understand only imperfectly.

In the following sections, I identify different types and prominent examples of such usage issues among kinesiology students in both speech and writing. Word processors sometimes—and in some cases, *always*—fail to flag these problems. Plus, such resources are not always available, as when we write a test longhand or give a speech on the fly.

TRICKY NONWORD/WORD PAIRS

It can be embarrassing to use what you think is a perfectly good word only to be told that it is not a word at all. This happens more often than you might think. Some nonwords stand out more in speech than on the page, and vice versa. Either way, however, all nonwords should be avoided, even when audiences or readers do not notice. Sometimes a professor or employer will notice, and they many be critical of you due to your mistake.

- alot ☒
- a lot ☑
    - Think of *a lot* as you would of *a ton*, or not just *a lot* but *a whole lot*. How silly would *aton* or *awholelot* be?
- heighth ☒
- height ☑
    - I surmise the nonword *heighth* quite likely derives from desired coherence with *length* and *width*. Why not *th* for all three?
- interpretate ☒
- interpret ☑
    - An interpretation comes from interpreting, certainly not *interpreat*ating. This is disanalogous to pairs like *communication/communicate*.
- its' ☒
- its ☑
- it's ☑
    - This may be a by-product of realizing that *it's* is not a possessive but a contraction. The rule is no apostrophe for possessive pronouns.
- momento ☒
- memento ☑
    - Think of mementos as triggering *mem*ories rather than capturing *mom*ents, though it is often moments we remember.
- pronunciate ☒
- pronounce ☑
    - As with *interpretate*, the mistake is that of back-forming the verb from the noun. Think verb/noun pairs like *explain/explanation*.
- smoothe ☒
- smooth ☑
    - Likely this stems from a false analogy with *soothe*, since smoothing is often soothing, yet the verb and the adjective are both spelled *smooth*.
- supposably ☒
- supposedly ☑
    - This no doubt comes from the idea that the object could *be* supposed (as in *supposable*) rather than one that *is* supposed.

- unequivocably ☒
- unequivocally ☑
    - Think of the root adjective *unequivocal*, which makes sense; *unequivocable* does not, as anything *could* be made indefinite.

TOP 20 CONFUSIONS

Kinesiology students can do themselves a real service in their writing and speaking by avoiding the following confusions, most of which cause professors like me, and some prospective clients, to cringe, if only inwardly.

- amount/number
    - Avoid *amount* for something countable, even if the actual counting would be tedious.
        - The amount of fans is overwhelming. ☒
        - The number of fans is overwhelming. ☑
        - The amount of cheering is overwhelming. ☑
- being/as
    - Avoid *being* where *as* will do, usually followed by noun phrases. Use *being* before adjectives, however.
        - Being a kinesiology student, I study biomechanics. ☒
        - As a kinesiology student, I study biomechanics. ☑
        - Being fair, the professor penalized all late essays. ☑
- between/among
    - Use *between* for twosomes, *among* for three or more.
        - Keep this between you, me, and the coach. ☒
        - Among you, me, and the coach, it was the right call. ☑
        - Between you and me, it was the right call. ☑
- can/may
    - *Can* implies mere possibility, which usually does not need to be asked ("Can I sit down?" "I don't know, *can* you?"), whereas *may* also suggests permission, which often does need to be asked.
        - Can we talk about the lab assignment? ☒
        - May we talk about the lab assignment? ☑
- comprise/compose
    - The exceedingly common mistake is to use *comprise* when one means *compose* or *constitute*. The parts do not comprise but rather compose the whole. It is the whole that comprises the parts.
        - Eight weeks of study comprise the course. ☒
        - The course comprises eight weeks of study. ☑

- each other / one another
  - As with *between* and *among*, use *each other* for exactly two, *one another* for three or more.
    - The football players high-fived each other. ☒
    - The football players high-fived one another. ☑
- effect/affect
  - The typical error is using the verb *effect* (bring about) when one means *affect* (influence). This is likely what motivated the meteoric rise of *impact* as a verb, which is also best avoided.
    - Exercise effects health. ☒
    - Exercise affects health. ☑
    - Exercise impacts health. ☒
- further/farther
  - Use *further* to indicate extent and *farther* to indicate distance.
    - I will run further than yesterday. ☒
    - I will run farther than yesterday. ☑
    - I will think further about my run. ☑
  - So does one stretch *further* or *farther*? Either! *Further* suggests range of motion, and *farther* suggests greater distance. One stretches both further *into* full extension but farther *along* a tape measure.[2]
- good/well
  - The typical error here is using *good* as an adverb, though clearly one can do good as well as doing well.
    - You did good by setting a personal best. ☒
    - You did well by setting a personal best. ☑
    - You did good by exhibiting sportsmanship. ☑
- her/she (him/he, me/I)
  - Avoid the common informal practice of substituting object pronouns for subject pronouns. Similarly, avoid subject pronouns after prepositions like *between*.
    - Her and me played mixed doubles. ☒
    - She and I played mixed doubles. ☑
    - Between you and I, the lineup will change. ☒
    - Between you and me, the lineup will change. ☑
- i.e./e.g.
  - Use *i.e.* (meaning "that is") to indicate a brief explanation. Use *e.g.* for one or more examples (think of the *eg* in "egg-sample"). The most common error here is using *i.e.* for examples.
    - Performance-enhancing drugs (i.e., steroids) are widely banned. ☒

    - Performance-enhancing drugs (e.g., steroids) are widely banned. ☑
- it's/its
  - The temptation is to use the apostrophe to indicate possession, as with nouns, but possessive pronouns lack apostrophes. Furthermore, *it's* is the contraction of *it is* and so will often be considered inappropriate in formal writing.
    - The team is at it's best. ☒
    - The team is at its best. ☑
- lay/lie (present)
  - Almost everyone now uses *lay* intransitively (i.e., without an object), though it is still strictly bad form, probably because no one wants to be thought of as *lying* in any sense, even lying *down*. In the present tense, *lay* needs an object.
    - After practice, lay down for a while. ☒
    - After practice, lie down for a while. ☑
    - After practice, lay the glove on the bench. ☑
- less/fewer
  - Avoid the tendency to use *less* for units or countable things. The exception here is where the units specify a benchmark above an implicit reference.
    - This meal plan allows less calories per day. ☒
    - This meal plan allows fewer calories per day. ☑
    - I consumed less [energy] than 2000 calories. ☑
    - I ran the mile in less [time] than five minutes. ☑
- literally/figuratively
  - Avoid using *literally* for emphasis, since, taken strictly, *literally* means "is true," which is something figurative language *never* is. Better to use either an emphatic tone (in speech) or, better still, more precise language.
    - When the coach benched me, I literally died. ☒
    - When the coach benched me, I *died.* ☑
    - When the coach benched me, I was devastated. ☑
- then/than
  - Often confused in student writing, *then* implies either sequence (A first, then B) or consequence (if A, then B), whereas *than* implies comparison.
    - My teammates are faster then me. ☒
    - My teammates are faster than me. ☑
    - I went home, then to practice. ☑

- there/their (there's/theirs)
  - *There* indicates place; *their* indicates possession. For the *there's/theirs* distinction, see *it's/its* above.
    - The coaches went their yesterday. ✗
    - The coaches went there yesterday. ✓
    - The coaches made their way there. ✓
- to/too
  - The usual mistake is to omit one *o* from *too*.
    - The seniors went to the Axe, and I did to. ✗
    - The seniors went to the Axe, and I did too. ✓
- who/whom
  - Avoid informal use of *who* as an object, especially at the start of a sentence. It helps to keep the pronoun next to the preposition, if there is a preposition. The reader might reject prepositions as words to end sentences with.
    - Who did you give the data to? ✗
    - Whom did you give the data to? ✓
    - To whom did you give the data? ✓
- who's/whose
  - Similar to *it's* and *there's* above, one may be tempted to use the apostrophe to indicate the possessive, though again, possessive pronouns do not take apostrophes. *Who's* is the contraction of *who is*. (Treat *you're/your* similarly.)
    - Who's gear is this? ✗
    - Whose gear is this? ✓

TRICKY SINGULAR/PLURAL PAIRS

In academia, there are many words in—or derived from—Greek or Latin that can trip up unwary students and professors alike, especially in speaking, when they fumble for the correct word or pronunciation, but also in writing. We will focus on some of the more common singular/plural pairs you should master. You may notice certain patterns.

- alumna/alumnae
  - A female graduate is an alumn*a*. Several are alumn*ae* (pronounced "alum-*nay*"). The gender-neutral alum is acceptable but slangy.
- alumnus/alumni
  - An alumn*us* is a male graduate. Several—male or mixed—are alumn*i*. A female is *not* an alumnus, nor is a single graduate *ever* an alumn*i*.

- analysis/analyses
  - The plural form of *analysis* is *analyses* (pronounced "analy-*sees*").
- criterion/criteria
  - Several criteria, one criteri*on* (pronounced "criteri-*un*"). Avoid the mistake of using *criteria* as a singular term, as in "That is an accepted criteria."
- datum/data
  - Several data, one datum or data point. Although *data* was originally strictly plural, it is now commonly used as singular shorthand for *data set.*[3]
- emeritus/emeriti
  - A retired professor might earn the honorary title of Professor Emeritus (or Emerita). Two or more are Professors Emeriti.
- hypothesis/hypotheses
  - A hypothesis is, roughly, an educated guess subject to testing. The plural is *hypotheses* (pronounced "hypothe-*sees*").
- medium/media
  - One medium, several media. Television is a medium. TV, radio, and print are collectively media. Best avoid using *media* as a singular term.
- phenomenon/phenomena
  - The most common error here is using *phenomena* as a singular term. Avoid using *phenomenon* as a plural or just adding an *s*.
- syllabus/syllabi
  - Professors sometimes refer to a course outline as a syllabus: one syllabus, two or more syllabi.
- synthesis/syntheses
  - A synthesis is a combination of two or more different elements. The corresponding plural is *syntheses* (pronounced "synth-e-*sees*").
- thesis/theses
  - This term can refer to an essay's central claim (thesis statement) or a type of scholarly work. The plural is *theses* (pronounced "the-*sees*").

12 MORE CONFUSIONS

In my experience, the usage errors addressed in the "Top 20 Confusions" section above are those that are most common among kinesiology students and that are vital for students to minimize. There are, however, other errors to guard against that crop up occasionally. Below are 12 more usage errors to watch out for.

- bad/badly
  - Students sometimes use the adverb as an adjective. One feels *bad* about doing *badly*. One only feels bad*ly* by not feeling enough.
    - I feel badly for missing the tryout. ✗
    - I feel bad for missing the tryout. ✓
- casual/causal
  - A *causal* connection (as in cause and effect) can be serious, anything but *casual*. Inattentive typists (typo-ists?) sometimes fail to notice the difference.
    - There is a casual link between nutrition and health. ✗
    - There is a causal link between nutrition and health. ✓
- compliment/complement
  - Avoid *compliment* (praise) when intending *complement* (pair with). Yin and yang are compl*e*mentary, not compl*i*mentary!
    - The forwards compliment one another. ✗
    - The forwards complement one another. ✓
- continuous/continual
  - The common error is to use *continuous* (uninterrupted) instead of *continual* (repeated).
    - She works out continuously all season. ✗
    - She works out continually all season. ✓
- defiantly/definitely
  - No one confuses the two meanings here. *Defiantly* often appears as a product of typo-ist inattention.
    - The hypothesis is defiantly plausible. ✗
    - The hypothesis is definitely plausible. ✓
- disinterested/uninterested
  - *Disinterested* means impartial, unbiased. One who is unengaged or bored is not *dis*interested but *un*interested.
    - The fight crowd was disinterested in the undercard. ✗
    - The fight crowd was uninterested in the undercard. ✓
    - The fight judges should be disinterested. ✓
- irregardless
  - Never use this botched offspring—synonymous with both its parents—of *regardless* and *irrespective*. Choose one of the parents instead.
    - Irregardless of your work ethic, you will lose. ✗
    - Regardless of your work ethic, you will lose. ✓
    - Irrespective of your work ethic, you will lose. ✓

- lay/laid (past)
  - *Lay* is intransitive, the past tense of *lie*. *Laid* is transitive, the past tense of *lay*. Compare with *lay/lie* above.
    - Yesterday I laid on the bench. ✗
    - Yesterday I lay on the bench. ✓
    - Yesterday I laid the glove on the bench. ✓
- portray/express
  - A portrayal is a *deliberate* representation. Expressions are representations that may or may not be deliberate.
    - The gold medalist's tears portrayed pure joy. ✗
    - The gold medalist's tears expressed pure joy. ✓
- principal/principle
  - As a noun, *principal* means highest institutional authority (as in schools), not to be confused with *principle*, meaning rule or tenet.
    - The player was cut as a matter of principal. ✗
    - The player was cut as a matter of principle. ✓
- rationale/rational
  - The noun *rationale* (i.e., motivation) should never be confused with the adjective *rational* (i.e., stands to reason). Not all rationales are rational.
    - The sport ethicist's argument was rationale. ✗
    - The sport ethicist's argument was rational. ✓
    - The sport ethicist's rationale was irrational. ✓
- regime/regimen
  - A regime is an authoritarian government, whereas a regimen is a planned pattern of behaviour or routine.
    - Our team has a new exercise regime. ✗
    - Our team has a new exercise regimen. ✓

TRICKY WORD/PHRASE PAIRS

Among the various pitfalls of texting culture is losing track in formal writing of whether to have a space between words that sometimes go together. This has become less intuitive recently, and it is helpful to keep certain examples and principles in mind. Not all examples are covered here, but you should get a sense of the issues involved in such choices generally.

- awhile / a while
    - There is no space in the adverb (*awhile*) when immediately following a verb, but there is a space in the indefinite noun (*a while*) following a preposition.
        - We should study for awhile. ☒
        - We should study awhile. ☑
        - We should study for a while. ☑
- anymore / any more
    - Use *anymore* as an equivalent to *any longer* (indicating time from now on). Use *any more* not for time but rather amount.
        - I shouldn't have anymore sports drink. ☒
        - I shouldn't have any more sports drink. ☑
        - I shouldn't have sports drink anymore. ☑
- anyway / any way
    - Use *anyway* to mean "besides" to change topics, and use *any way* to suggest one among different ways.
        - Anyway we want to play is fine. ☒
        - Any way we want to play is fine. ☑
        - Any way, we want to play. ☒
        - Anyway, we want to play. ☑
- everyday / every day
    - Avoid misusing the adjective (*everyday*: common, typical) as an adverb (*every day*: daily) and vice versa.
        - I take a walk everyday. ☒
        - I take a walk every day. ☑
        - Everyday activities include walking. ☑
- onstage / on stage
    - Use *onstage* as an adjective, and use *on stage* as an adverb. (Treat *inline*, *onboard*, *upfront*, etc., similarly to *everyday*.)[4]
        - The dancer pirouetted onstage. ☒
        - The dancer pirouetted on stage. ☑
        - The onstage pirouette was flawless. ☑
- sometime / some time
    - There is no space in indefinite nonspecific times, but there is a space in indefinite specific times after a preposition. (Treat *someday / some day* similarly.)
        - We should train some time. ☒
        - We should train sometime. ☑
        - We should train at some time. ☑

- workout / work out
    - Use *workout* as a noun or adjective. The verb is *work out*.
        - I workout six days a week. ✗
        - I work out six days a week. ✓
        - This work out is challenging. ✗
        - This workout is challenging. ✓
        - This is a challenging workout routine. ✓
    - The same is also true in many similar cases (e.g., *breakdown*, *faceoff*, *knockout*, *makeup*, *setup*, *standby*, *standup*, *strikeout*, *takedown*, and *tryout*, which should be treated similarly to *workout*). Removing the space between verb and preposition often yields a noun or adjective.[5]

Apart from formal contexts, such minutiae of good form often seem, and often are, unimportant. Paying attention to them in formal contexts, however, can be part of what separates great communicators from good, good from merely competent, and competent from bad. You may be surprised at how little genuine effort it takes to climb, rung by rung, up this ladder.

BRIEF RECAP

In this chapter, we covered many common usage tendencies to avoid in formal communication:

- Informal versus formal usage
- Tricky nonword/word pairs
- Top 20 confusions
- Tricky singular/plural pairs
- 12 more confusions
- Tricky word/phrase pairs

CHAPTER 8

Grammar Guidance

CHAPTER OVERVIEW

You already have a fairly good, if implicit, grasp of grammar, although there are some principles worth making explicit to improve both your understanding of them and your facility with them. We begin by characterizing grammar rules in general, proceeding then to the issue of sentences that either lack or have too many parts. We follow this by distinguishing among different types of agreement (not just subject-verb agreement), then grammatical ambiguity and various issues I refer to as category mistakes. We finish off by considering how rules of grammar are not wholly distinct from but perhaps surprisingly overlap with rules of usage, punctuation, and style.

GRAMMATICAL RULES

Principles of usage concern how to use words correctly. Principles of composition concern how to string sentences together correctly. In between lie principles of grammar, which concern how to string words together into sentences. Though being a native English speaker implies that you have a reliable understanding of many principles of grammar, that understanding remains largely implicit and is often overburdened by demands of scholarly and professional communication. As such, it is worthwhile to make some of this intuitive understanding more explicit and alert you to common grammatical gaffes and corresponding solutions.

Words are not only meaningful but also structurally poised. Their meaning will constrain what they are used to do, but their structural properties will constrain how they may be combined with other words. For instance, *mat* is a noun indicating a type of thing that can go under other types of things. The world has many different individual mats and varieties of mats, and yet very often we use *mat* to refer to a particular one using an article: *the* to mean a specific, definite mat or *a* to mean a specific, indefinite one. The definite article *the* allows us to use the word *mat* to refer to one specific, definite mat. The preposition *on* refers to a relation between two different things, whereas *under* usually indicates the same relation in reverse. The word *is* is a versatile verb indicating predication, identity, or existence itself.

Consider the task of forming a complete sentence using all and only the following words (allowing for capitalization and punctuation later): *the*, *the*, *cat*, *mat*, *is*, and *on*. These words are structurally poised in that only some combinations will yield complete sentences. The most likely combination is "The cat is on the mat." Other meanings can also be generated, while some strings will be meaningless, violating rules of grammar:

- The cat is on the mat. (target meaning)
- The mat is on the cat. (different meaning)
- The the cat mat is on. (meaningless)

As constraining as grammatical rules are, however, these structural constraints allow for varied syntax and thus a certain degree of freedom even in this tightly constrained example. Compare:

- The cat is on the mat. (target meaning)
- On the mat is the cat. (same meaning)
- On the mat the cat is. (same meaning)

Here we contrast a "normal" way of expressing the thought with an old-fashioned, almost poetic form and, finally, what I call **Yoda syntax**: saving the verb for the end of the sentence. This is a somewhat odd but still well-formed structure in English, yet it is more common in other languages, such as Latin.

The rules of grammar change over time, and there is an element of arbitrariness in many decisions about which rules are in effect and how strictly they are to be applied. To take just one example, you have probably been told by a teacher at some point that a preposition is something you should *never* end a sentence with. I just violated that rule: *with* is a preposition. But some sentence constructions are awkward unless they end in prepositions (e.g., the longstanding joke phrase, variously attributed, "something up with which I will not put"). Likewise, you may have observed that Shakespeare himself often violates this "rule," which only stresses the changeable and somewhat arbitrary nature of certain grammar rules. Some professors are absolute sticklers about sentence-ending prepositions, whereas others, like me, are not.[1]

As you read on, think of your grammar knowledge as good but needing to become a little more explicit and refined, with the constraints of grammar rules being important but less limiting of variety than you may have previously assumed.

SENTENCES WITH MISSING OR EXTRA PARTS

We can think of a sentence as a string of words obeying grammatical rules and containing both a **subject** and a **predicate**. In this sense, the subject is what the sentence is about, and the predicate tells us something about it. The subject is a noun or something that functions like one, a pronoun or description. The predicate contains at least a verb and often other parts of speech: an adjective, adverb, object, and so on. In the simple sentence "I play," for instance, the subject is the pronoun *I*, and the predicate is the verb *play*. In "I play hockey," the subject is still *I* but the predicate is now "play hockey," including the verb *play* and the verb's object, *hockey*. If, instead, I write "The person who wrote *Communication for Kinesiology* plays," the predicate is only *plays*, and the subject is the definite description "The person who wrote *Communication for Kinesiology*"—the subject is not *just* "The person," and the verb *wrote* is *not* part of the predicate, but is rather part of the grammatical subject.

Sentence Fragments

A **sentence fragment** is an incomplete sentence written as if it were complete, with a capital at the beginning and a period at the end. Students tend to write sentence fragments deliberately to emphasize the fragment content or accidentally because they have lost sight of the grammatical structure of an unwieldy description or dependent clause. Most sentence fragments do not belong in formal writing. Some may be formally permissible, however, when part of certain interactive devices to engage the reader.

Some examples will help. We begin with a typical deliberate sentence fragment together with an easy fix. Note that although "I love" could stand as a sentence on its own, adding *which* results in a **dependent clause**, which could not be its own sentence. If, by contrast, a string of words could serve as a stand-alone sentence, it is an **independent clause**, even if contained in a larger sentence.

- I play hockey. Which I love. ✗
- I play hockey, which I love. ✓

This sort of choppy writing, again, is often motivated by a desire to emphasize the fragment by setting it off. This trend may be seen in various types of popular writing, including literary fiction. Notwithstanding the literary liberty, however, in scholarly and professional writing, such emphasis should be either removed or otherwise revised:

- Many Canadians play hockey. Passionately. ✗
- Many Canadians play hockey passionately. ✓

- Many Canadians play hockey *passionately*. ✓
- Many Canadians play hockey—passionately. ✓

Although deliberate sentence fragments are usually best avoided, there are, even in more formal writing, exceptions. As Shertzer (1986) observes, "If the subject or predicate is not expressed, it must be readily understood from sentences that precede or follow" (p. 6). A rhetorical question, for example, may prompt a permissibly fragmentary reply:

- How do many Canadians play hockey? Passionately. ✓
- Are Canadians passionate about hockey? Indeed. ✓

It may seem somewhat arbitrary to allow fragmentary answers to rhetorical questions but advise against fragmentary emphasis. That is the nature of linguistic and formal conventions, which are often arbitrary but should be respected just as the artificial rules of any sport or game are respected.[2]

Harder to spot for many students is the **run-on fragment**, where the grammar of either a subject, predicate, or dependent clause is complicated enough to be confused for a full sentence. In such cases, the student has been either careless or too ambitious. The solution is twofold: first, to be more attentive, and second, to correct the grammar by rewriting, omitting what is inessential or adding what is essential, often from the preceding or following sentence:

- The first time a person with a sedentary lifestyle attempts an exercise program and the initial results can be disappointing. ✗
- The first time a person with a sedentary lifestyle attempts an exercise program, the initial results can be disappointing. ✓
- The first time a person with a sedentary lifestyle attempts an exercise program and the initial results are disappointing, it is crucial to encourage them to keep at it. ✓
- Although some ethicists have derived arguments from liberal and libertarian principles that favour lifting bans on performance-enhancing drugs. ✗
- Some ethicists have derived arguments from liberal and libertarian principles that favour lifting bans on performance-enhancing drugs. ✓
- Many people oppose legalizing steroids, although some ethicists have derived arguments from liberal and libertarian principles that favour lifting bans on performance-enhancing drugs. ✓

Sentence Fragment Tips

- Avoid when emphasizing content
- Beware of confusing for sentences, especially
    - Long descriptions
    - Long dependent clauses
- Ways of correcting
    - Insert missing content
    - Attach appropriately to preceding or subsequent sentences

Run-On Sentences

Whereas a sentence fragment is missing parts and is not a sentence at all, a **run-on sentence** has extra parts. Some perfectly well-formed sentences can be grammatically complex and challenging to readers, and some instructors may mark these as run-on sentences as a matter of pragmatic or stylistic preference rather than strict grammatical complaint. It is a good rule of thumb to aim for less than 50 words for most sentences, although this is hardly a strict rule. The more crucial type of run-on sentence to watch for stitches two or more independent clauses together either without punctuation or with a comma (a so-called comma splice). Recognizing the problem is key. There are two easy ways to fix it. First, try each independent clause as its own sentence. If the sentence pair seems too choppy in context, try joining the clauses into a **loose sentence** with a comma and conjunction.[3] Take the above pair, "Recognizing the problem is key. There are two easy ways to fix it," which students might be tempted to write as a run-on sentence with or without a splicing comma. Once the problem is identified, the corrections are straightforward:

- Recognizing the problem is key there are two easy ways to fix it. ✗
- Recognizing the problem is key, there are two easy ways to fix it. ✗
- Recognizing the problem is key. There are two easy ways to fix it. ✓
- Recognizing the problem is key, and there are two easy ways to fix it. ✓

Run-On Sentences

- Rule of thumb: less than 50 words for most sentences
- Finding run-ons: look for sentences with more than one independent clause
- Rewriting run-ons
    - As unique sentences (first option)
    - As a loose sentence (second option)

AGREEMENT

There are different ways in which the parts of a sentence can fail to fit together grammatically. I include here not only what grammarians call **agreement** but also related "failures of fit" that, taken together, give an abstract perspective on a more general problem.

Probably the most common grammatical disagreement in student writing is that between subject and verb. This typically happens with complex subject terms where the subject's number (singular or plural) is mistaken. If the subject term is simply *employers*, for instance, as in "Employers ____ looking for candidates with diverse skill sets," students would almost never fill in the blank with *is* instead of *are*. With more a complex subject, such as "Employers in the field of kinesiology," however, students are apt to mistake the singularity of *kinesiology* as indicating a singular subject, when in fact the subject is still, as in the simple case, plural:

- Employers in the field of kinesiology *is* looking for candidates with diverse skill sets. ✗
- Employers in the field of kinesiology *are* looking for candidates with diverse skill sets. ✓

One of the reasons students may miss such errors is that word processing grammar checks often miss them.

A similar problem occurs with collective nouns. We often think of groups, implicitly and reductively, in terms of their individual members. However, collective nouns should be treated as singular:

- When equipment gets sufficiently worn, *they* should be replaced. ✗
- When equipment gets sufficiently worn, *it* should be replaced. ✓

Just as the subject of a sentence must agree with the verb, so too must the subject "agree" with the predicate:

- The purpose of this essay *defends* the view that esports should be considered genuine sports. ✗
- The purpose of this essay *is to defend* the view that esports should be considered genuine sports.[4] ✓

Another related concern is ensuring that parallel phrasing is used when possible. The "agreement" here is between words or phrases that can and should be expressed in similar form:

- Forwards and *those on defence* run different drills. ☒
- Forwards and *defenders* run different drills. ☑
- Those on offence and *those on defence* run different drills. ☑

Notice that we are straddling the distinction between considerations of grammar and considerations of style. Though grammar and style are largely separate concerns, at some point the good form of grammar and the good form of style will inevitably coincide. This overlap will also be seen at grammar's respective borders with usage and punctuation.

Types of Agreement

- Subject-verb (beware complex subjects)
- Pronoun-noun
- Subject-predicate
- Parallel phrasing

AMBIGUITY

Ambiguous grammar lends itself to different interpretations of a sentence. Whereas **vagueness** is a matter of language that lacks precise meaning, **ambiguity** is a matter of language that has different possible precise meanings. If I say that a certain conditioning program is *okay*, that will, in many formal contexts, be too vague. Vagueness is often a product of imprecise word choice. Ambiguity, on the other hand, can be a product of word choice or grammar—or even, in some cases, a blend of the two.

In the case of ambiguous pronouns, more than one noun or noun phrase may plausibly be read as substituting for it. In the below example, for instance, which is the plausible thing: the theory or the objection? Correcting the problem requires disambiguation, which means either ensuring there is only one possible noun that the pronoun could be substituted for or specifying the referent without use of a pronoun:

- The theory faces an objection. *It* is plausible. ☒
- Such theories face an objection. *It* is plausible. ☑
- The theory faces an objection. *The objection* is plausible. ☑

A purer case of grammatical ambiguity can arise in describing past events. In describing two related events in the simple past, for instance, the relation will often be lost. In our example below, for instance, it is not clear whether the content of the past imagined thought was that the restocking was finished or that it was in process:

- She imagined that he *restocked* the lab supplies. ✗
- She imagined that he *had restocked* the lab supplies. ✓
- She imagined that he *was restocking* the lab supplies. ✓

Similar, if subtler, ambiguities can arise from shifting tenses or moods mid-sentence. The editorial correction is to either rewrite one of the verbs for consistency or, if the sentence allows, eliminate one of the verbs.

- This stretch will increase range of motion and *does* not predispose tissues to injury. ✗
- This stretch will increase range of motion and *will* not predispose tissues to injury. ✓
- This stretch will increase range of motion and not predispose tissues to injury.[5] ✓

It should be noted, however, that a tense shift will sometimes be appropriate to clarify different temporal relations to the text:

- Jesse Owens was a great athlete, and his gold medal performances *will be* remembered for many years to come. ✓

Yet another type of ambiguity arises from modifiers that "squint." A modifier such as the word *only* becomes a **squinting modifier** when used ambiguously. In the uncorrected example below, it is unclear whether the intended meaning is that fall-term play is limited to weekends or that weekend play is limited to the fall term. Disambiguation either way solves the problem:

- Our team plays on weekends only during fall term. ✗
- Our team plays only on weekends during fall term. ✓
- Our team plays on weekends during fall term only. ✓

Types of Ambiguity

- Pronoun reference
- Temporal relations
- Tense or mood shift
- Squinting modifiers

CATEGORY MISTAKES

Several significant grammatical problems can be grouped under the heading **category mistakes**, where one implies, usually unintentionally, that something belongs in a certain category when in fact it does not.[6] Catching such errors requires that you pay closer attention to the implicit logic of the language used.

Take modifier problems. A **dangling modifier** occurs when the proper subject is implicit and the explicit subject is logically inappropriate. A **misplaced modifier** is meant to modify one word or phrase but is placed too far away and thus modifies another. In the examples below, we are accidentally implying, in turn, that a playing field literally ran drills and that a coach wanted their team to play hard during halftime rather than during the game.

- Dangling modifier
  - <u>Running drills</u>, *the field* was in great shape.[7] ✗
  - <u>Running drills</u>, *we* saw that the field was in great shape. ✓
- Misplaced modifier
  - The coach <u>spoke</u> about playing hard *during halftime.* ✗
  - The coach <u>spoke</u> *during halftime* about playing hard. ✓

A related kind of category mistake is implying logically inappropriate comparisons. The correction in such cases is to rewrite the descriptive language so that it fits with both categories. In the uncorrected example below, for instance, we imply that Holt *and* his writing belong to the category of dull *professors* rather than dull *things*:

- Holt is a <u>dull professor</u> and so is *his writing.* ✗
- Holt is <u>dull</u> and so is *his writing.* ✓

Some category mistakes are common in conditional or "if-then" sentences, especially in informal speech. A **conditional sentence** has an if-then structure linking two propositions: If P1, then P2. In the conditional "If you study, then you will pass," the first proposition "you study" is called the **antecedent**, and the second, "you will pass," is the **consequent**. Sometimes the *then* is implied, as in "If you study, you will pass." Conditional verbs such as *would* or *could* tend to be logically appropriate only in the consequent, although informal speech often includes them in the antecedent:

- If you <u>would have</u> studied, you *would have* passed. ✗
- If you <u>had</u> studied, you *would have* passed.[8] ✓

Whereas conditional antecedents are common in informal contexts and are frowned upon in formal ones, a recent popular trend in formal contexts is using **prepositional gestures**, which suggest spatial imagery, but are logically nebulous if not inapt.[9] In formal contexts, we will often hear of discussion *around* a problem, not *about* the problem, or of working *toward* a solution, not *for* a solution, of difficulties *surrounding* a project, not *with* a project, and the like. Despite growing formal acceptance of such usage, the terminology is weak, imprecise, and a form of bureaucratese, and so should be discouraged. Spatial prepositions should usually be reserved for descriptions of spatial relations:

- The faculty discussion was *around* standardizing methods. ✗
- The faculty discussion was *about* standardizing methods. ✓

Types of Category Mistakes

- Modifiers
    - Dangling
    - Misplaced
- Inapt comparisons
- Conditional antecedents
- Prepositional gestures

GRAMMAR AND OTHER DOMAINS

As with many other linguistic rules, the rules of grammar have partly arbitrary origins but, once established, should be respected, at least in formal contexts. Certain instances above may be better described in terms of usage, punctuation, or style rather than grammar. Ambiguous pronouns, for instance, indicate flawed grammar and usage. Sentence fragments exhibit flaws in grammar and punctuation. A lack of parallel phrasing is a grammatical and stylistic flaw. In this sense, though analyzed separately, these domains of linguistic concern are better integrated than our analysis might suggest. This will become clear in the next chapter in the case of punctuation.

BRIEF RECAP

In this chapter, we examined different issues relating to grammar:

- Grammatical rules
- Sentences with missing or extra parts
- Agreement
- Ambiguity
- Category mistakes
- Grammar and other domains

CHAPTER 9

Punctuation Pointers

CHAPTER OVERVIEW

In an age of instant messaging, the convenience of the quick and sloppy text may make it hard to see, much less grasp, the finer points of punctuation. In this chapter, we begin with a call to attend more closely to punctuation than we do, or should, in informal contexts. Then we turn to the subtleties of using commas, quotation marks, and apostrophes. After this, we distinguish a trio of punctuation pairs that are easy to confuse: colons and semicolons, parentheses and brackets, and hyphens and dashes (including en dashes and em dashes). We end the chapter with a look at the instructive case of "epiquotation" and the possibility of creating new forms of punctuation.

PROPER PUNCTUATION

As with the rules of usage and grammar, the rules governing punctuation are, in a way, arbitrary and yet should be respected in formal contexts. The arbitrary basis does not imply that we may simply ignore their normative force. In a way, then, punctuation may be seen to reflect, for instance, when we might pause in uttering a sentence. This might be clearest in the case of commas and periods. Other punctuation marks have a similar function, setting off and clarifying relations among different linguistic items.

It is crucial that you pay closer attention to proper punctuation in formal writing than you would when text messaging or even writing informal emails. Indeed, the prevalence of these communicative practices makes it harder for many of us to remember how to punctuate correctly or notice when we have gone astray. Although what follows will not be a completely remedial or crash course in punctuation, it will be a refresher of certain key points to remember, as well as perhaps extending your know-how a little.

COMMAS

You should think of commas as both linking and separating different linguistic items, very often either elements in a series or different clauses. The series could contain as few as two adjectives or three items in a list. As for clauses, dependent clauses should be linked by a comma, whereas independent clauses may be linked by a comma, together with a coordinating conjunction, into a loose sentence. Examples will clarify.

First, where two or more adjectives modify the same noun or phrase, they should usually be "link-separated" by a comma if not linked by a conjunction:

- It was a long hard-fought game. ✗
- It was a long, hard-fought game. ✓
- It was a long and hard-fought game. ✓

For a listed series, two items are usually linked with a conjunction, whereas three or more should be link-separated with commas:

- We need new gear and a deeper bench. ✓
- We need new gear a deeper bench and better leadership. ✗
- We need new gear, a deeper bench, and better leadership. ✓

You may have noticed that there are two different conventions for using commas in lists, where one includes and the other excludes a comma following the second-last item, just ahead of the conjunction. Both conventions are correct, so whichever you follow, the most important thing is to be consistent. Sticking with the above example, consider the following:

- We need new gear, a deeper bench and better leadership. ✓
- We need new gear, a deeper bench, and better leadership. ✓

The comma before the conjunction is sometimes called a serial, series, or Oxford comma. Given the conventional status of both approaches, it might seem foolish to have a principled preference here.[1] However, I, among others, recommend the serial comma, as it adds a measure of clarity to most lists without any real cost. As Dreyer (2019) puts it, "No sentence has been harmed by a series comma, and many a sentence has been improved by one" (p. 24).

Commas should also be used to link dependent clauses to independent clauses, especially at the end of sentences but also at the beginning with longer dependent clauses or when omitting the comma would cause confusion. Note in the second example below how one tends to read the comma-less version as a sentence fragment.

- I play hockey which I love. ✗
- I play hockey, which I love. ✓
- Though I had been injured in the previous game I played. ✗
- Though I had been injured in the previous game, I played. ✓

But resist the temptation to link separate independent clauses in this way, as this yields a **comma splice**, which is bad form. To correct comma splices, rewrite the clauses as separate sentences or add a coordinating conjunction after the comma:

- The cat is on the mat, the dog is on the log. ✗
- The cat is on the mat. The dog is on the log. ✓
- The cat is on the mat, and the dog is on the log. ✓

Use what is called the *vocative* comma before a quotation *unless* the preceding language is either an independent clause or fits seamlessly with the grammar of the quotation. In the case of an independent clause, use a colon instead of a comma. (In-text citations are needed here but have been omitted for clarity.)

- Holt states "some figure skating performances count as both sport and art." ✗
- Holt states, "some figure skating performances count as both sport and art." ✓
- Holt states the following conclusion: "some figure skating performances count as both sport and art." ✓
- Holt states that "some figure skating performances count as both sport and art." ✓

Commas may be used in pairs to bookend asides, content that is inessential to a sentence but worth adding for informational and stylistic reasons. The typical error in this case is omitting one of the two commas. Think instead that commas here are operating as parentheses, or "bookends," coming in pairs:

- Barthes, a French theorist wrote the book *What Is Sport?* ✗
- Barthes a French theorist, wrote the book *What Is Sport?* ✗
- Barthes, a French theorist, wrote the book *What Is Sport?* ✓

Among various other uses of the comma, I choose to end this section on an amusing note concerning the "only" comma, which is used to indicate uniqueness. Here is an example Dreyer (2019) refers to as the "best illustration of the necessity of the 'only' comma I've ever managed to rustle up" (p. 33). The comma indicates that Elizabeth Taylor had only one second marriage, whereas the lack of a comma indicates more than one marriage to the same person:

- Elizabeth Taylor's second marriage, to Michael Wilding ☑
- Elizabeth Taylor's second marriage to Richard Burton[2] ☑

Comma Tips

- Use series commas.
- Use commas with dependent clauses.
- Avoid comma splices.
- Use the vocative comma to introduce a quotation unless
    - the introducing language is an independent clause
    - the quotation grammar fits the introducing language grammar
- Use a pair of commas to bookend asides.

QUOTATION MARKS

The principal use of quotation marks in writing is to identify specific language spoken or written elsewhere. In fiction, we are to imagine a character speaking words placed inside quotation marks. In scholarly or professional writing, we identify material we get from another source, usually accompanied by a citation of the source. Besides this basic use, which is familiar to most students, several more are worth identifying and clarifying. These include mentioning language, meta-quotation, identifying short work titles, and scare quotes.

Whereas quotation marks are normally used to indicate the specific words someone spoke or wrote on a certain occasion, they can also serve the related function of mentioning language in general. In such cases, the use of double quotation marks is standard, but consider the alternative of using italics instead, as in the convention followed in this book:

- The word kinesiology derives from Greek. ☒
- The word "kinesiology" derives from Greek. ☑
- The word *kinesiology* derives from Greek. ☑

This applies when double quotation marks are standard for quoted material, which is not always the case.[3] Single quotation marks should be used to indicate matter quoted or mentioned within a quoted passage, that is, a **meta-quotation**:

- "Hemphill uses the term "cybersport" to designate "alternative sport realities"…" ☒
- "Hemphill uses the term 'cybersport' to designate 'alternative sport realities'…"[4] ☑

Quotation marks should also be used to indicate the titles of short works. Long works are, by contrast, normally given in italics.

- The article "The Philosopher's Stallone" was published in the journal *Cinematic Codes Review.*
- The chapter "Sport Card Portraiture" is part of the anthology *Portraits and Philosophy.*
- The song "Even When You're Miles Away" appears on the album *Would That It Were.*
- The poem "Seventeen Syllables" is included in the collection *Poems for Another Time.*

Quotation marks can also be used to suggest loose usage, often attributed by the writer to other people. These are called **scare quotes** and are appropriate whenever you would use an "air quotes" gesture in conversation. Avoid using quotation marks for added emphasis since this will create the wrong impression. Use italics instead, as they suggest that you *really* mean it, whereas scare quotes suggest the opposite.[5]

- Some people champion the "sports" of chess and videogames. ☑
- Some people champion the "non-sports" of chess and videogames. ☒
- Some people champion the *non-sports* of chess and videogames. ☑

Quotation Marks

- Double
  - Mentioned language
  - Quoted material
  - Short work titles
  - Loose usage
- Single
  - Quotes within quotes

APOSTROPHES

Apostrophes have two primary functions: indicating possession or contraction, respectively. Students can find proper use of the apostrophe challenging, especially when dealing with either possessive forms of names or other nouns ending in *s*, possessive pronouns, or plurals.

Form possessives of singular nouns by adding an *'s* to the end of the noun, even if that noun ends in *s*. Although you will sometimes see the *s* omitted where the name ends in an *s* and the pronunciation of the added *'s* seems awkward, follow this rule of thumb unless instructed otherwise. At any rate, you can always *pronounce* the *s's* possessive with a silent second *s*.

- Socrates' approach to education... ✗
- Socrates's approach to education... ✓

With plural possessives, by contrast, if the plural term ends in an *s*, add only an apostrophe after the *s*. A common mistake people make is to add an *'s* to the singular form as if doing this forms a plural possessive. Where the plural term does not end in an *s*, add an *'s* as if it were singular. In the example below, because *bacteria* is already plural, its possessive needs an *s*, but *not* in order to pluralize it.[6]

- The Holt's only child... ✗
- The Holts' only child... ✓
- The bacteria' resistance... ✗
- The bacterias' resistance... ✗
- The bacteria's resistance... ✓

One source of confusion is the fact that, unlike other possessives, possessive pronouns do *not* take an apostrophe. The common mistake here is to treat possessive pronouns as if they were singular or plural possessive nouns:

- Every dog has it's day. ✗
- Every dog has its day. ✓
- Forwards run their drills, defenders their's. ✗
- Forwards run their drills, defenders theirs'. ✗
- Forwards run their drills, defenders theirs. ✓

Never use an *'s* to pluralize words. This is a common error, sometimes stemming from laziness or inattention. An *'s may* be used to pluralize *nonwords only*, as with acronyms. Even with acronyms, however, the conventions are shifting. In general, you should pluralize nonwords *without* an apostrophe *unless* the result would be confusing to read:

- This theory faces several objection's. ✗
- This theory faces several objections. ✓
- The fight is available on several DVD's. ✓
- The fight is available on several DVDs. ✓
- I earned four A-s last semester. ✗
- I earned four A-'s last semester. ✓

We turn now to contractions, which are often discouraged in formal writing for being unnecessarily informal. This rule is not absolute, though. There are two kinds of exceptions to be aware of. Some texts contain contractions, including classic works and more current informal sources. Quoting such sources requires including the contractions, of course, but even alluding to them may be preferable with contractions rather than without. Suppose we want to quote from or allude to the famous line from *Hamlet* "Whether 'tis nobler in the mind to suffer the slings and arrows of outrageous fortune."[7] You must include the *'tis* rather than *it is* in a quotation. In a subtler reference, say, in the title of a research paper on overuse injuries, the allusion of "'Tis not in fact nobler" will be more accessible than "It is not in fact nobler," which would likely elude the reader altogether.[8] Likewise, in certain disciplines there is now greater tolerance of occasional contractions when the corresponding uncontracted form would be somewhat stilted. Even when a formal context seems to tolerate contractions, however, they are usually best avoided.

Apostrophe Use

- Possessives
  - Singular: add an apostrophe + *s*
  - Plural (ending in s): add apostrophe *after* the *s*
  - Plural (not ending in s): add apostrophe + *s*
  - Pronouns: add *no* apostrophe
- Plurals
  - Nonwords *only*, especially if hard to read otherwise
- Contractions
  - In quoted material
  - Best avoided, even where formal context allows

NOT TO BE CONFUSED

We turn now to a few pairs of punctuation marks that are sometimes confused. These confusable pairs have different functions, and keeping these different functions in mind may help guide you in deploying them properly. Along the way, I will highlight some common traps writers can fall into, especially at the student level.

Colons and Semicolons

Colons and semicolons are similar to periods in linking independent clauses to further linguistic content. In function, however, they differ both from periods and from each other. They suggest a closer relationship between what precedes and what follows.

Some uses of the colon tend to be straightforward, such as separating titles from subtitles (e.g., *Communication for Kinesiology: A Practical Guide*) and hours from minutes in references to time (e.g., 3:15 pm). Often when a colon should be used to link independent clauses to further content, however, a writer errs by using either a period or a semicolon. Such further content may be, for example, an explanation, quotation, or list. A colon in such cases signifies that the writer will specify what has been announced in the independent clause. Neither a period nor a semicolon announces that further specification is coming:

- Athletic success depends on two virtues. Courage and hard work. ✗
- Athletic success depends on two virtues; courage and hard work. ✗
- Athletic success depends on two virtues: courage and hard work. ✓

In some contexts, semicolon use is unavoidable. For instance, documentation styles may require semicolons to separate sources in in-text citations (e.g., [Jones, 2021; Smith, 2020]). As well, semicolons should be used in place of commas when, for instance, one or more items in a list contains a comma itself. In this helpful example from Dreyer (2019, p. 43), using commas rather than semicolons makes it seem as if we are talking about five or six books, not three:

- Lucy's favorite novels are *Raise High the Roof Beam, Carpenters, Farewell, My Lovely,* and *One Time, One Place.* ✗
- Lucy's favorite novels are *Raise High the Roof Beam, Carpenters*; *Farewell, My Lovely;* and *One Time, One Place.* ✓

Semicolons may also be used to suggest an implicit connection between one independent clause and another, but in such cases a period would do just as well on its own or, better, with an explanation of that implicit connection. Likewise, when the connection is explicit, a period tends to be no less effective:

- Bob barely passed biomechanics; he had failed to keep up all term. ✓
- Bob barely passed biomechanics. He had failed to keep up all term. ✓
- Bob barely passed biomechanics as he had failed to keep up all term. ✓
- The initial results were promising; however, the test was ultimately inconclusive. ✓
- The initial results were promising. However, the test was ultimately inconclusive. ✓

My somewhat controversial view, then, is that you should avoid semicolon use unless absolutely necessary, and it almost never is. The first decade or so of my professorial career involved many published articles and even a few books and, by my count, *zero* semicolons![9]

Colon and Semicolon Use

- Colons
  - Use after independent clauses to link specifying content
- Semicolons
  - Include if necessary (e.g., documentation, disambiguation)
  - Avoid if possible (e.g., where a colon or period would do)

Parentheses and Brackets

Parentheses are sometimes called "brackets," but these terms strictly designate different types of punctuation. Parentheses are the familiar "round brackets" used for asides (such as this one) and for in-text citations in many documentation styles (e.g., APA). True **brackets**, on the other hand, are "square parentheses." Their proper use is less common and less well-grasped.

When a parenthetical remark on its own is an entire sentence or more, the period is placed *before* the close parenthesis, not after:

- (This aside is a complete sentence). ☒
- (This aside is a complete sentence.) ☑

Two common uses of brackets are for in-quote interpolations and for parentheses within parentheses (i.e., "**meta-parentheses**"). In the case of in-quote interpolations, sometimes you will improve the flow of your writing by inserting clarifying content not provided in the original text. In such a case, the brackets mean that the quoter has added to or replaced something in the original. Either way, it is vital to preserve the intended meaning of the original. Note how, removed from its original context, the quotation below leaves *its* entirely nebulous. Likewise, we can indicate by putting the *T* in brackets that the independent clause quoted did not begin with *To* (though some styles consider this scrupulous attention excessive):

- "But to discover that the character of the England captain was even better than we thought enhances its aesthetics even further." ☒
- "[T]o discover that the character of the England captain was even better than we thought enhances its [i.e., the game's] aesthetics even further." ☑

- "[T]o discover that the character of the England captain was even better than we thought enhances [the game's] aesthetics even further."[10] ☑

Think of meta-parentheses as analogous to meta-quotations. Just as single quotation marks should be used for quotations embedded within quotations, so should brackets be used for parentheses embedded in parenthetical remarks:

- (as is argued in Smith (2001)) ☒
- (as is argued in Smith [2001]) ☑

As with quotation marks, both parentheses and brackets should come in pairs. Any time you have an open parenthesis, *(*, or bracket, *[*, you should be looking to include its corresponding close parenthesis, *)*, or bracket, *]*, and vice versa. As obvious as such advice seems, you might be surprised just how often careless editing flouts it.

Parentheses and Brackets

- Parentheses
    - In-text asides
    - In-text citations
- Brackets
    - In-quote interpolations
    - Meta-parentheses

Hyphens and Dashes

Hyphens and dashes are related and often used indiscriminately, although they each have proper uses. First, we need to distinguish the hyphen from the **en dash** and **em dash** respectively:

- Hyphen: - (type a hyphen)
- En dash: – (type any letter, a space, two hyphens, a space, any letter, and another space)
- Em dash: — (type any letter, two hyphens, any letter, and a space)

The en and em dashes are so-called because they are roughly the width of an *N* and *M* in turn. The principal uses of hyphens and dashes are as follows: hyphens for compound adjectives, en dashes for margins and ranges, and em dashes—not individual hyphens—for setting off emphatic asides:

- *Hyphen (compound adjectives)*
  - It was a hard-fought game.
  - That was a one-in-a-million shot.
- *En dash (margins and ranges)*
  - Acadia won the game 7–1.
  - See pages 1–2 in the manual.
- *Em dash (emphatic asides)*
  - Mixed martial arts—the best sport invented in my lifetime—is televised on some sports networks.
  - Team sports have an enhanced aesthetic appeal—though why this is so is unclear.

Besides such basic uses, a few complications are noteworthy. First, some documentation styles tolerate, explicitly or implicitly, the use of hyphens rather than en dashes for ranges. APA is an example. Relatedly, some word processing programs do not make it easy to create en or em dashes. In such cases, hyphens may have to suffice for ranges, whereas em dashes may be typed as double hyphens (--) but *never* as single hyphens. You may also notice that some publications use spaced en dashes rather than unspaced em dashes for setting off asides. Last, en dashes may serve a hyphen-like function for compound adjectives involving phrases and not just individual words. A hyphen works to create a compound adjective from *Gretzky* and *style*, although not from *Wayne Gretzky* and *style*, as the hyphen includes *Gretzky* but not *Wayne*:

- It was a Gretzky-style play. ☑
- It was a Wayne Gretzky-style play. ☒
- It was a Wayne Gretzky–style play. ☑

Hyphen and Dash Use

- Hyphens
  - For compound adjectives
  - Replace with en dashes (for phrase compounds)
- En dashes
  - For margins and ranges
  - Replaceable with hyphens (if necessary)
- Em dashes
  - For emphatic asides
  - Replaceable with double hyphens (*not* single)

"EPIQUOTATION"

To remind us of the evolving nature of punctuation, which follows mutable conventions in service of certain functional roles, I will end by briefly considering a punctuational solution to a certain problem posed by the cultural phenomenon that I call "epiquotation." An **epiquotation** is a line that misquotes a source—usually a popular work of fiction—and is widely known to be in error but persists in a culture because of its aesthetic appeal (Holt, 2016a). For example, "Luke, I am your father" is not a line from *The Empire Strikes Back*. The actual line is "No, *I* am your father." Similarly, "Alas, poor Yorick! I knew him well" is not a line from *Hamlet*. It is rather "Alas, poor Yorick! I knew him, Horatio."[11]

My coining of *epiquotation* for this phenomenon was motivated partly by the fact that, in my judgment at least, current terminology and punctuation fail to capture this phenomenon. It is not *just* a (mis)quotation, an interpolation, or a paraphrase. That is why I propose, in addition to this new term, a new punctuation mark to go with it (Holt, 2016a), one that captures both the unified and the approximate nature of epiquotations. Since quotation marks often look like an equal sign (=) rotated ninety degrees, I thought epiquotation marks should be an *approximately* equal sign (≈) rotated ninety degrees. So we may speak of epiquoting Darth Vader or epiquoting from *The Empire Strikes Back*:

- "Luke, I am your father." (quotation) ✕
- "[Luke,] I am your father." (interpolation) ✕
- Luke, I am your father. (paraphrase) ✕
- ≈Luke, I am your father.≈ (epiquotation) ✓

Figure 9.1. Epiquotation Marks
Source: Jason Holt

As with usage and grammar, the moral is to see *through* the conventions of punctuation to the function they serve. The point of the point is the point.

BRIEF RECAP

In this chapter, we covered some of the finer points of punctuation:

- Proper punctuation
- Commas
- Quotation marks
- Apostrophes
- Not to be confused punctuation pairs
- "Epiquotation"

CHAPTER 10

Style Matters

CHAPTER OVERVIEW

Just as dressing with style for a formal occasion can be an asset, so too can writing with style. In both cases, you aim to put your best foot forward. In writing, however, it is not yourself but your *thoughts* that you aim to present well. In this chapter, we cover some essential matters of writing style, starting off with cohesion, then coherence, then moving on to various style tips: balancing focus on language and thought, altering sentence length and structure, and using specific devices to engage the reader.

HAVING STYLE

Students sometimes make the mistake of supposing that good writing is only a matter of getting the content right and leave aside questions of style. There are different reasons for this. One is the assumption that what makes good writing good is what it says and not how it says it, its content and not its form, its substance and not its style. Students often worry so much about getting good content into their writing that they see style as something extra they have neither the time nor the inclination to focus on. You might be disappointed if a professor tells you your writing lacks cohesion, coherence, or "style" in some general sense. You might be frustrated too. The feedback may be appropriate, but on its own it tends to be unhelpful, leaving you with at best a vague sense of what the problem is and at worst virtually no guidance to solve it.

The assumption that style is dispensable in writing is common but wrong. Students should be concerned with style and not just substance.[1] Distinguishing form and content is easy enough, but that does not mean they can be separated so easily or evaluated completely independently. Think of the care you would take getting ready for a date or a job interview. You might be a catch or a perfect fit for the position, but you still want to put your best foot forward. Making an effort to improve your writing style is as important as getting the content right. Failing to do so is noticeable. There are common patterns of unstylish writing among students. In what follows, I identify several specific elements of style characteristically missing in poorer student writing, along with concrete suggestions for how to improve in those areas. When it comes to style matters, style matters.

COHESION

Perhaps the vaguest-seeming complaint about student writing is that the writing somehow "just doesn't flow." Unhelpful as this might seem on its own, the notion of a sense of flow can be identified with a passage's *cohesion* (Williams & Colomb, 2012, p. 36). The meaning of the word *cohesion* is less than obvious, though in fact, the term's meaning can be made precise. Once we grasp what cohesion is, the lack or presence of it in a passage can be specified concretely.

Here is a definition: **cohesion** is the sense of flow created by linking adjacent sentences. This definition seems simple enough, but it has three parts, and each part is important. The first part—the sense of flow—gives a way of telling, in the experience of reading itself, whether some passage has or lacks cohesion. A reader who experiences this sense of flow as they read knows that the passage has cohesion. But a reader who fails to experience this sense of flow, who finds the writing choppy and the transitions abrupt or awkward, knows that the passage lacks cohesion.

Definition of Cohesion

Cohesion is the sense of flow created by linking adjacent sentences.

The second part of the definition—the creation of links—indicates the mechanism that a writer has used, deliberately or unconsciously, to create that sense of flow. This mechanism will be specified using an example below. The third part of the definition—adjacent sentences—tells you where to look for the mechanism that gives the passage its flow, if it has any. So the first part of the definition gives a criterion, a way of telling from experience whether a passage is cohesive. The second part gives the mechanism of cohesion, and the third part gives the location. That location is specifically between the end of one sentence and the beginning of the very next sentence.

Here is an example. Suppose you are writing an essay in sports history and your draft contains the following two sentences, each giving a relevant piece of information about Wayne Gretzky:

- Wayne Gretzky was traded by the Oilers in 1988. Wayne Gretzky became a hockey icon.

Students tend to do a decent job in getting such information into their writing. There are two bits of information about the same subject: Gretzky himself. Even so, however, this pair of sentences is not at all cohesive. Because you have used the full name in both sentences, it is not even clear to a reader that you have remembered that both statements concern the same individual. Such choppy writing reveals the laziness of pasting one piece of information to another and is typical of student writing that could use a style makeover.

One thing you might do to make the sentence pair more cohesive is to use, instead of the full name in the second sentence, just the surname *Gretzky*, or, even better, the pronoun *He*.[2] You could write one of the following pairs:

- Wayne Gretzky was traded by the Oilers in 1988. *Gretzky* became a hockey icon.
- Wayne Gretzky was traded by the Oilers in 1988. *He* became a hockey icon.

You have now made it clear that both bits of information are about the same subject and that you know this to be true. The writing flows a little better, but neither sentence pair is what we would call particularly cohesive.

According to Williams and Colomb (2012, p. 38), for a sentence pair to be cohesive, the beginning of the second sentence should refer to information contained in the first sentence. This is true up to a point. But in the revised examples above, the use of the surname *Gretzky* without the first name, and even more so, the use of the pronoun *He*, makes it clear that both pieces of information are about the same person. You can add significantly to the cohesion in this sentence pair, however—and this goes for sentence pairs generally—with a more specific mechanism: using linking language between the beginning of the second sentence and the *end* of the first. That makes for a truly cohesive sentence pair.

Consider the related point that there are some things a reader will not be able to tell about the two pieces of information contained in our sentence pair. One thing is the historical order of the two events mentioned, specifically Gretzky's trade and his achievement of hockey icon status. Did he become an icon only after the trade? At the time of the trade? Or did iconic status come first? The reader may know, and may also suspect that the writer knows, but it would be better if the writer could demonstrate this knowledge to the reader.

Most kinesiology students, and certainly most hockey fans, know that Gretzky's status as a hockey icon was secured before the trade from the Oilers in 1988. Consider now the following revision to our pair of sentences:

- Wayne Gretzky was traded by the Oilers in 1988. *By then, he had become* a hockey icon.

You can see here how we have created a sense of flow in this pair of sentences. The "By then" that starts the second sentence refers to the event mentioned at the end of the first sentence, the trade in 1988. Not only that, but by using "had become" we have also implied the additional content about how the two pieces of information are historically related: Gretzky became an icon before the trade.

This illustrates the important point that style can help substance along and is something students should pay attention to in drafting and revising their writing. Keeping in mind that your professor wants to know not just that you know pieces of information but, more importantly, that you see relationships among different pieces of information can help you insert in your writing the kind of linking language between sentence pairs that creates cohesion.

Pro Tip: Creating Cohesion

Create cohesion between two sentences by linking the *beginning* of the second sentence to the *end* of the first sentence.

COHERENCE

Another complaint that students may hear about their writing is that it lacks **coherence**. Though we sometimes speak of language as *in*coherent when it fails to make sense *at all*, the substance of the complaint here is rather different. Usually, a piece of writing that your professor calls incoherent *does* make sense, sentence by sentence at any rate, or even paragraph by paragraph, but the writer has failed to integrate the different parts of the text well.

Coherence contrasts nicely with cohesion, which we discussed above, the latter shedding light on the former. Cohesion concerns how *one* single part fits with the *next* single part, how one sentence links to the one following, how that links to the one following it, and so forth. By contrast, coherence is not about how a single part fits with the next, but rather how *all* the parts fit together. This distinction can be usefully illustrated with the following example:

- Everybody likes movies. I watched a lot as a kid. Children have different needs, which are stronger than desires. Freud thought that desires originate in the subconscious mind.

You probably think that because there is something clearly wrong with this passage, it fails to exhibit cohesion. That is intuitive but mistaken. Remember, cohesion means that the start of one sentence refers to, or links in some way with, the end of the sentence immediately before it.

In the above passage, you will notice that each sentence pair actually is cohesive, in the following way. Where the first sentence ends with a reference to movies, the second sentence begins with a statement about watching movies. Where the second sentence ends with the word *kid*, the following sentence begins with the plural *children*, a synonym for *kids*. Likewise, that sentence ends with a reference to desires, and the following sentence begins to detail what Freud thought about such states of mind. Is this passage cohesive? You bet.

What it lacks, however, is a different virtue of good style: coherence. The problem with the passage is not that it fails to exhibit a sense of flow, because it does have flow. The problem is that the flow takes us on a meandering rather than a straight course. One sentence flows into the next, but the writer does not stay on track, and the topic is hard to identify. In fact, there are several topics raised in this passage, and none of them is obviously the main topic. The key to having a coherent passage, then, is to have a *unified* topic. In this passage, we move from everybody liking movies to the writer saying that they watched a lot as a kid, and from there to children's needs in general versus their desires, and finally ending with Freud's theory of where desires come from. Again, we move from one topic to another relatively easily, and yet nothing unifies this passage at all.

The key to having a unified topic in a given paragraph is, obviously, to stay on point. In formal writing, the best way to ensure that a paragraph is unified by a single topic is to deliberately include a topic sentence. A topic sentence not only tells the reader what the subject of the paragraph is but also makes a claim that you intend to support, illustrate, explain, or otherwise develop in all the other sentences in the paragraph.

Let us turn now to an example of a paragraph with a clear topic sentence that provides structure and unity to the entire passage. The passage also illustrates, conveniently, how linking language between adjacent sentences creates cohesion. So what we have here is a case where the two virtues—cohesion and coherence—coincide:

> There are various unproblematic senses in which sports and games are virtual or 'not real'. On Huizinga's account of play, for instance, and considering them species of play, these activities stand 'outside "ordinary" life' (1950, 13). Similarly, on Suits's theory of games, players accept the artificial constraints of rules not found in everyday life because it is those rules that define a game (1973, 55). As such, sports and games are different from what may be seen as the reality of the utilitarian workaday world, no less valuable for that reason. Sports like kendo and fencing are derived from, and virtual representations of, martial sword-fighting, and American football, rugby, and other team sports are often metaphorically described in terms of war, without being confused for it. (Holt, 2016b, p. 5)[3]

If you read this passage carefully, you should be able to tell that the sentence tying the whole paragraph together, the topic sentence, is the very first one. This is not uncommon, to find the topic sentence right at the start of a paragraph, although this is not always where you will find topic sentences, nor are there always topic sentences in paragraphs. The first sentence concerns sports and games being "unreal," and each of the other sentences explores different aspects of this general claim, thus providing support for the topic sentence by partly illustrating and partly explaining it.

Pro Tip: Creating Coherence

Create paragraph coherence by sticking to one subject indicated by a topic sentence.

So the passage is coherent. But if you look carefully at the adjacent sentence pairs, you should find linking language that makes the passage cohesive as well. Linking the first sentence to the second is the aside phrase "for instance," which could have appeared right at the beginning of that second sentence. The first word in the third sentence likewise indicates to readers that the content of the statement will be similar to that of the second sentence, a point about games being unreal. At the beginning of the fourth sentence, we have the phrase "As such." This indicates a transition between those two previous sentences and the rather complex set of content that takes up the rest of this last sentence in the paragraph. The final sentence, beginning with *Sports*, is in effect a continuation of the *sports* after *as such* in the preceding sentence, as though the latter covers both. In other words, given what has just been said about play and games, we list particular examples of sports illustrating that part of the topic sentence itself.

If you aim for a balance in your writing between cohesion and coherence, you stand a good chance of producing writing with good style. But besides these general considerations, I turn now to specific tips and strategies for taking your writing style to the next level.

STYLE TIPS

As we saw with principles of composition, what follows are specific suggestions and guidelines rather than absolute rules for improving your writing style. Some of these suggestions are about specific changes to your attitude as a writer, and others are concrete elements to use in your writing. The style tips have been pared down to a half dozen essentials falling into three categories: balancing focus, altering length and structure, and using devices. We can further compress these essentials into the phrase "balance, alteration, devices" or, even better, the mnemonic BAD. Remember the mnemonic. For good style, go BAD.

B for Balance

Life is all about balance, but so, in many respects, is having good style, in writing or in anything else. Of the different ways in which good writing style requires balance, two in particular are worth highlighting.

Try balancing your focus on the thoughts you hope to express in your writing and on the language itself. Students can be susceptible to overlooking their writing by overfocusing on their thoughts, often under the misimpression that it is the thoughts alone that count, with "right-feeling" words automatically fitting them. They assume, in other words, that their thoughts can be transparently read *through* the words themselves. This is seriously distorted thinking. The most brilliant ideas in the world, badly expressed in language, are just so much bad language.

Some students are, on the contrary, not concerned enough with the thoughts they try to express in the words they use, wanting just to fill up enough of the page—or enough pages—to complete the assignment at hand. Still, the more common and significant problem is that of not paying enough attention to the specific language used, and so more focus on *the language itself*, and not just on what they are trying to express, would serve many students well. Ultimately, the goal is to have balanced focus on both the language used and the thoughts expressed.

Good style also requires not just cohesion and coherence, as we saw above, but a balance of emphasis between the two virtues. Many students do well enough writing paragraphs that are substantially coherent, but often to the neglect of cohesion, yielding choppy and disjointed writing despite thematic unity. The complementary, though less common, problem is sometimes evident as well. Students prone to unreflective, automatic writing can fall into this trap, writing a text so quickly that it may flow well—in a kind of stream-of-consciousness way—but be incoherent. If students keep in mind both virtues, they stand a much better chance of not losing sight of either in the pursuit of the other.

A for Alteration

People tend to fall into habits, as we know, which can easily turn into ruts, which are hard to get out of. This is true as much in writing as in other aspects of life and applies no less to great writers than to mediocre ones. One way to improve your writing style is to deliberately alter certain features of the sentences you write. Two deserve mention here: length and structure.

Vary the length of your sentences: some shorter, some longer, some your usual length. In most cases, you should aim for variations within a range of about 10 to 50 words, which in 12-point Times New Roman font should fall between about half a line to about three full lines of text. This is not to say that you should always avoid shorter or longer sentences than specified in this range. Sometimes short, punchy sentences, no more than a few words, or even one, are called for. In other cases, especially if your grammar skills are honed, continuing beyond three lines is perfectly acceptable. If you find yourself writing a sentence longer than about 50 words, though, make sure you have not already made a mistake—like creating a run-on sentence or comma splice—and consider whether two shorter sentences would be better. Most of your sentences should fall into the 15- to 35-word range. What is the minimum number of words a sentence can have? One. What is the maximum? In principle, it is infinite, although practical considerations for readers and writers alike will limit sentence length well below this threshold. Bear in mind that longer sentences increase the chance of uncaught errors.

Along with varying sentence length, you should aim to alter sentence structure, in aid of avoiding the type of rut we spoke of above. Many of us tend to write simple sentences like this one. It can, however, help to vary the structure by, for instance, including asides as I just did twice. By contrast, beginning a sentence with a dependent clause, which I just did, can also help. Useful, too, can be adding further content or comment with a dependent *which* clause, which I am doing right now in this very sentence. The more you change up your style, the more likely you are to avoid falling into the wrong kind of rut.

D for Devices

Perhaps the most concrete suggestion I have is for you to use rather specific mechanical devices in your writing, only some of which will be discussed here. Your writing can be more engaging if you add conversational devices and, in some cases, even wordplay.

Conversational devices can help engage the reader by positioning them to feel as if they are part of a dialogue with you. Can I give an example? One is using rhetorical questions, where you ask a question and then answer it yourself, as I just did. Another is adding verbs that position the reader as their implicit subject. Consider my beginning this sentence with *consider*. This makes the text more like a conversation, which tends to engage readers more. In some subdisciplines in kinesiology, it is acceptable common practice to explicitly address the reader as *you*, oneself as *I*, or both together as *we*. This is the case in the philosophy of sport, where first- and second-person language is often used, as are occasional contractions. You also may have noticed that in this book I often directly refer to you as *you* and myself as *I*. We hazard some danger in edging too close to informal usage and tone, especially in some subdisciplines. You can expect such usage to get red-penciled in exercise physiology or biomechanics. However, I would insist that the payoff is often worth the risk, especially where students minimize the risk through being attentive to the norms in whatever subdiscipline they are working.

Wordplay includes coming up with catchy phrases, puns (i.e., plays on words), and even poetic qualities like rhythm. Particularly good writing style may have certain rhythmic patterns, as with unrhyming song lyrics. There can be subtler cases of wordplay where writers choose some words and phrases that sound better, even if the reason is somewhat elusive. Consider why the phrase "poor sport" sounds better than "bad sport." This is because *poor* and *sport* are linked by a common "or" sound unmatched by *bad* and *sport*.[4] The use of catchy phrases, puns, and rhymes can be particularly effective in titles, section headings, and lines at the beginning or end of an essay or other major piece of writing to encapsulate in slogan form the view you want to convey.[5] Such wordplay may entertain the reader as well as demonstrating your creativity. When your creative juices are really flowing, in rare cases it might even be worth coining a new word.[6] What if I combined *wordplay* with *playful* to coin *wordplayful*, for instance?[7] The use of devices that noticeably emphasize the *play* in wordplay should be used sparingly, but they can be quite effective tools for styling student prose.

Brief Style Tips: Breaking "BAD"

- *B*alance of attention
  - Language and thought
  - Cohesion and coherence
- *A*lteration of sentence features
  - Length and structure
- *D*evices to use
  - Conversational and wordplay

BRIEF RECAP

In this chapter, we examined several issues concerning writing style:

- Having style
- Cohesion
- Coherence
- Style tips

CHAPTER 11

Giving Presentations

CHAPTER OVERVIEW

In this chapter, we turn to what, for many, is a cause of significant anxiety: giving presentations. We begin by identifying different types of speech distinguished by delivery and purpose. We then look at engaging the audience: employing behavioural and structural techniques to help catch and keep audience attention. Next comes advice for preparing yourself to present, comporting yourself as you present, and avoiding common delivery problems. Last, we draw similarities and distinctions between in-person and online presentations.

PUBLIC SPEAKING

A lot of this book is either concerned explicitly with writing or no doubt has been interpreted by you as applying primarily to writing. This impression, though false, is certainly understandable. After all, most of us, in most situations, are fairly comfortable speaking to someone else or with small groups. Writing is not everyone's strength, nor is it in everyone's comfort zone. Although some of us might be uncomfortable introducing ourselves to strangers, for the most part we are more comfortable speaking than writing. The exception, of course, is *public* speaking, which many of us would rather *not* do even if we are reasonably good at it. Writing is safer, we think. At least it feels safer. Talking one on one or in small groups feels safer too. Like it or not, though, public speaking is an inevitable part of academic study and is indispensable in many of the professions associated with your kinesiology degree.

TYPES OF SPEECH

We often refer to public speaking in scholarly and professional contexts as *giving a presentation* rather than *making a speech*. There is a corresponding tendency to underemphasize the fact that, even when visual aids accompany your talk, the primary vehicle of information transmission and meaning sharing is the speech itself. By "speech," though, I mean not just spoken language, even though that is typical of speechmaking. Most speeches are given vocally, to be sure, but giving a speech to a group of people who are hearing impaired might be done through sign language.

Types of speeches can be distinguished in various ways. For our purposes, it will be helpful to distinguish among these types in terms of, respectively, their mode of delivery and underlying main purpose. A **set speech** is based on a prewritten script. The script can be read by the speaker directly from a printed copy or teleprompter. In the case of, say, actors on stage or in movies, the speech will be memorized by the actor, for instance, when whoever is playing Hamlet in *Hamlet* recites that famous soliloquy. An **impromptu speech** is the opposite, not only unscripted but not even prepared for, totally off the cuff. It might seem hard to think of an example, but when your professor answers an unexpected question in class, that counts as an impromptu speech, even though the lecture itself was probably well prepared.

The virtue of the set speech is that the speechwriter is able to craft the message carefully and so does not have to worry about coming up with wording on the fly. The downside, though, is that set speeches are fixed and often seem awkward or stilted when delivered, especially when students read from a script in class, which makes maintaining sufficient eye contact a substantial challenge. On the other hand, the virtue of the impromptu style is spontaneity and flexibility that does not take away from audience engagement. The corresponding downside here, of course, is not preparing material to be delivered. An unprepared in-class presentation is liable to be bad unless the professor forces an impromptu speech for learning purposes.

The third type distinguished by mode of delivery is the **extemporaneous speech**, which, like a set speech, is prepared but, unlike a set speech, is unscripted.[1] The extemporaneous form, like the impromptu, tends to facilitate audience engagement, flexibility, and spontaneity, although, unlike the impromptu, it allows for significant planning and preparation. In this way, the extemporaneous speech captures many virtues of both set and impromptu speeches while avoiding the pitfalls of either. If you have the choice of which presentation format to choose for your speech, then, I strongly recommend you opt for the extemporaneous, which blends the virtues of preparedness and interactivity.[2] Where other considerations outweigh this advice, let them prevail.

Types of speech can also be distinguished by purpose. Obviously, any given speech could serve a variety of different purposes, but in most formal speeches, one of the following purposes will very likely predominate (Rolls, 2007). An **informative speech** is intended to impart information to an audience. Most university lectures fall into this category. A **persuasive speech** is meant to change or galvanize people's beliefs or behaviour. Most political and motivational speeches fall into this category. Think also of a coach's motivational speech at halftime or even a researcher at a conference giving a presentation. A **demonstrative speech** is intended to teach the audience how to do something. Think, in this case, of a physical education teacher showing students how to perform some skill or a television chef walking viewers through one of their recipes. The **social speech**, by contrast, serves a function relative to a public event—commemorative, facilitating, or constitutive—as with eulogies, toasts, introducing guest speakers, and so on. Most speeches you give at the university level will fall into the informative or persuasive category, though the other speech types may become more prominent in your eventual profession. If you end up a physical education teacher, for instance, you will give a lot of demonstrative speeches in class, especially at lower grade levels.

Types of Speech

- Distinguished by *delivery*
    - Set (prepared, scripted)
        - Read
        - Memorized
    - Impromptu (unprepared, unscripted)
    - Extemporaneous (prepared, unscripted)
- Distinguished by *purpose*
    - Informative
    - Persuasive
    - Demonstrative
    - Social

In this section, I have focused on different kinds of speech assuming they are individual rather than group presentations. Group presentations present special difficulties in coordinating group members in both preparation and presentation phases. A fair division of labour is crucial for both phases, as is coordinating elements of the presentation beyond the demands of individual presentations. For example, if group members coordinate their attire, an individual speaker can dress appropriately without having to worry about their outfit clashing with the outfits of their co-presenters.

ENGAGING YOUR AUDIENCE

Public speakers often give insufficient attention to trying to engage the audience. For one thing, preparing and delivering material can take a lot of time and effort, with little room to worry about whether the presentation will go over well, except perhaps for marked presentations, where you probably care about the impression you make on the professor. It also makes sense to avoid caring too much about engaging the audience because that is, in effect, beyond your control and so a potential source of anxiety. Even so, it is wise to focus not on *whether* you engage your audience but on *what* you can do to make such engagement more likely. Engaging an audience involves drawing and keeping their attention and structuring the presentation content to help the audience receive it as well or as impartially as they can. If, for whatever reason, the audience is not already on your side, you want them to at least give you a fair hearing.

Audience Attention

Whether you are giving an in-class presentation or presenting research findings at a conference, it can be hard to get and keep an audience's attention. This applies whether you are presenting to classmates who know you well or to conference-goers who are complete strangers. Good speakers can struggle to engage an audience when they are—but even when they are not—just one in a slew of class or conference presenters. In such cases, you may not be able to make yourself stand out from the other speakers on the docket, but you can still take certain steps to lure and land the attention of your audience, meeting them halfway.

Part of the battle may be won, in some cases, by the circumstances in which you present. If there are concurrent sessions at your conference, for instance, then your audience, having had other options, has at least minimal interest in your topic or else is perhaps intrigued by your title. Similarly, professors tend to find it easier to teach electives, as that implies that students have at least a minimal preference for the course. Core courses, by contrast, are far more challenging to teach in a way that engages students, many of whom are there by necessity. Being effectively forced to listen to a speech tends to bias an audience against it. So when you have to give a presentation in a core class, your classmates' attention will likely be hard to get and harder to hold. At least it will often feel that way.

Since you, as a speaker, and others, as an audience, are prone to distractions, it will be helpful to do what you can to minimize distractions for all concerned. Turning off your smartphone is a no-brainer, and in most formal contexts, the convention is for audience members to avoid the use of a phone or even to avoid having it out or on (unless necessary for some reason).[3] More generally in this regard, Russell and Munter (2014) give a mnemonic acronym to remind speakers to help the audience to, as they put it, "*RESIST* distractions":

- *R*
    - Point out what's *relevant.*
        - Flag a key message.
        - Explain "what's in it for them" throughout your talk.
        - Offer a solution to a pressing problem.

- *E*
    - Make an *emotional* connection.
        - Share personal information that makes you likable.
        - Refer to something that makes you laugh.
        - Use photos or movie clips to capture a feeling.
- *S*
    - Add an element of *surprise.*
        - Include a counterintuitive or shocking statistic.
        - Make a noticeable change in your delivery style.
        - Share a fact about your topic that surprised you.
- *I*
    - Find a way to *include* people.
        - Ask people how to solve a relevant problem.
        - Let the audience choose what you will talk about next.
        - Get people to talk with someone sitting next to them.
- *S*
    - Engage their *senses.*
        - Add sensory language to examples or sound bites.
        - Use visuals (or videos) that features images (and motion).
        - Bring a prop that people can see, hear, smell, or touch.
- *T*
    - Create some *tension.*
        - Remain silent longer than seems comfortable.
        - Include a video clip that features conflict.
        - Make people vote on a controversial issue. (Russell & Munter, 2014, p. 40)[4]

Although it would be unrealistic to try to follow each piece of the above advice for every single presentation you give, most of the time, one or another subset of these recommendations should prove useful in planning a presentation.

On a cautionary note, there are counterproductive ways of getting audience attention. An unfortunate way of doing this is to dress inappropriately. To minimize distraction potential for both you and your audience, wear clothes and shoes that are comfortable and appropriate. If elements of your wardrobe (e.g., high heels) are uncomfortable, that will be a distraction to you, and plausibly for your audience as well. Clothing can be comfortable but too informal for giving presentations (e.g., athletic attire) or dressy but distracting (e.g., eveningwear). Informal clothes prime the audience to expect a mediocre presentation. Inappropriately dressy clothes appear to be a misfired attempt at professionalism.

Aim for the sweet spot in the middle. For in-class presentations, dress in business casual attire: collared shirts or blouses; slacks, khakis, or knee-length skirts (even dark jeans); and comfortable but stylish shoes (even dark sneakers). For conferences, business causal may be fine as well, depending on the subdiscipline, but the expectations may be higher, and business attire may be more appropriate in some cases: suits and pantsuits, or at least coat and tie or blazer, blouse, and skirt, with shoes to match. The take-home point is that the comfortable and sometimes revealing clothing you may often wear to class can be a liability when giving presentations, even in class. If an audience is paying more attention to your appearance than to your talk, then something has gone wrong.

Structural Engagement

Outlining a prepared formal presentation is no less important than outlining a scholarly research essay. The first step in outlining a speech should be, in broadest terms, what kind of approach to take. Often advisable is a **direct approach**, where you state the main point(s) both at the start of your presentation and, to reinforce and emphasize the point(s), again at the end. For an audience, this approach is easy to follow and relatively efficient, and it shows sensitivity to the audience's interest (Russell & Munter, 2014). You should note how scholarly presentations so structured will mirror the structure of scholarly essays: thesis stated at the beginning, justified in the main body, and then restated at the end. Indeed, many scholarly presentations are just that: presentations of essays.

Such speeches are easy to follow because, in a main point, the audience has a conceptual anchor for any support material given. They can appreciate relevant connections between pieces of information as steps on a path if they know the intended destination. Naturally, if the audience is unaware of where a speech is going, each part may appear random rather than an element in a coordinated whole. A speaker who hides the conclusion until the very end of the talk, as a kind of surprise, has exhibited the **indirect approach**. Despite the virtues of the direct approach, the indirect approach might be better if, for instance, your thesis is controversial, the audience is otherwise biased or highly critical, or the indirect approach is indicated by the parameters of the specific speech or the culture in which it is given (Russell & Munter, 2014). Still, although the idea of building the drama of your presentation to a big finish may appeal to your aesthetic sensibilities, this should be resisted unless you have an exceptionally strong reason to make an exception. To put this advice for formal presentations in slogan form: be direct if possible, indirect if necessary.

Direct or Indirect?

Be direct if possible, indirect if necessary.

Planning the structure of your presentation in more detail can also help you make it more engaging. For instance, you can model the beginning and end of your speech on the introduction and conclusion of a formal essay. Likewise, the bulk of your presentation might follow a similar pattern to that of the main body of an essay, with important information and arguments grouped into separate, logically ordered sections. As with essays, whether the grouping of material or sectioning comes first may vary, although the two will be mutually informative. The essential things at this stage are (1) to create information chunks and (2) to order the chunks in a way that make sense given your main point(s). For instance, if your topic is the conditioning program for a particular sport, it might make sense to chunk information into different types of sport-relevant conditioning—say, strength, aerobic capacity, and flexibility—in order of relative importance to the sport in question. If your topic is a process, say, how exercise affects metabolism, it might be best to structure your talk as the process itself is structured: initial conditions, causal factors, and then effects, perhaps short-term and long-term, individual then social. On the other hand, if your purpose is explanatory, it could be better to identify the phenomenon and work backwards to the explanation. With a problem-solution model, the explanation could be the middle step between a problem identified and a solution proposed.

DELIVERY TIPS AND PROBLEMS

Assuming the content is taken care of, we now raise questions of delivery. Preparing a speech is one thing, and delivering it is quite another. Many of the kinesiology students in my classes have some wherewithal giving speeches and tend to be better, in at least some respects, than many students I have taught in other areas. At the same time, these strengths may benefit from reinforcement and sometimes appear to be offset by corresponding weaknesses. I will begin with tips on preparing yourself for the presentation, then discuss a few key points to remember during the presentation itself, and finally finish off with what in my experience are the top five common flaws in kinesiology presentations, whether given by students or professionals.

Preparing Yourself

Preparing the material is essential, of course, but so, too, is preparing yourself to present. In other words, even if the material is ready to go, you may not be. Once the material is prepared, turn to these other preparatory matters.

1. *Know the essentials*. Think about the broad structure of your talk in point form, as if you had to fit it all on one index card. Indeed, writing down the basic point-form structure of your talk on an index card can be a

useful practice. You should know, without having to consult notes, the following: your opener, your take-home message(s), your main points of support, and your closer. Including your introduction and conclusion, in most cases you should aim to have about six bullet points—no fewer than five and no more than seven. Such essentials will often include the correct pronunciation of technical terms, researcher names, and so forth. Too many otherwise well-prepared talks can be hamstrung by students haltingly pronouncing terms or names they are used to seeing on paper but not saying out loud.

2. *Be question-ready.* This may seem counterintuitive, but it is a good idea to prepare more than you present. One reason is that cutting out some material will help you select better-quality material to include. Also, following your presentation, there will often be either a planned or impromptu question period, so it is good to be prepared for possible questions even if no Q&A is planned, because people may approach you afterward with follow-up questions. Preparing more than you present also—because it actually makes it so—tends to make you feel more knowledgeable about your topic. Obviously, you will present the most important material you have prepared. Having a little more information, however, such as background information on a key experiment or historical event, will make you feel better prepared to deliver the material you have planned to, as well as revealing just how knowledgeable you are should the occasion arise.
3. *Prepare support materials early.* By "support materials," I mean whatever notes you will consult while presenting and any visual aids you have planned. Preparing such materials well in advance will help you remain relatively relaxed as the presentation date approaches. It will allow you to practise giving your speech with the very same support materials you will use during the presentation itself. This can aid you psychologically in the preparation phase and in the delivery phase alike. It can also help you fine-tune your presentation material.
4. *Practise intelligently.* Depending on the type of speech you are scheduled to give, it may or may not be possible to do full practice runs. In the case of short speeches, obviously it will be more feasible to do so. In such cases, it may be a good idea to ask friends, family, or classmates to listen to a dry run and give you feedback. Asking a classmate to help in this regard can be useful for both parties, since you can listen to your classmate's speech in return for listening to yours, a matter of reciprocal goodwill. The most important point for intelligent practising is to replicate, to the extent you can, the circumstances of the actual presentation. So, having the notes you plan to consult on hand will help. Rewriting your notes just before presenting after you have practised with a rough set is a bad idea. Get the notes more or less finalized before you

practise. If you can practise in the same venue, great. That will be a bonus, but this is not always feasible. Even wearing the clothing you plan to wear when giving the speech can help cue your mind, when you are actually giving the speech, to the lessons learned in practice.

5. *Take care of yourself.* Giving a speech, even an in-class presentation that is worth only a few grade points, can be stressful. In such a context, of course, self-care is important. Do your best to get a good night's sleep the night before, eat a decent breakfast, and show up to the presentation venue a little early, taking a relaxed bathroom break, double-checking your presentation materials, and so forth. As for attire and grooming, you should aim for comfort.[5] You should avoid clothes that look good but add unnecessary discomfort to what is already plausibly a tense situation. Likewise, you should pay attention to, but not stress out about, your appearance. Your appearance does count, of course, but being too well groomed or overly distracted by a bad hair day will not help you get ready to present.

As You Present

There are many things you will focus on when giving a speech. Just keeping yourself going and on track, along with keeping track of key elements you will touch on, is hard enough. This alone often will seem too much to handle. Nonetheless, there are some delivery pointers worth keeping in mind during your speech. *What* you present in a speech is certainly important, though no less important is *how* you present it.

1. *Maintain balanced eye contact.* Some presenters make the mistake of never looking at the audience. Others make the opposite error, staring too long at one person, often a friend or someone who seems receptive to the speech. The first mistake is one of not engaging the audience. The second is one of excluding most of the audience. Don't be stingy with eye contact. Spread it around. In any case, avoid trying to maintain eye contact the whole time. It is hard not to be thrown off by the constant inundation of rich though ambiguous nonverbal feedback from the audience. Think of a "one-off-two-off-three-off" procedure where, while giving the speech, you look at one person for a few seconds, then look off at the wall—a good opportunity to formulate your thoughts—then shift your gaze to a new person, and repeat.
2. *Project your voice sufficiently—or slightly more.* Many of the kinesiology students in my classes manage to project sufficiently during a presentation. Still, not all do, and so the message to project one's voice sufficiently is worth reinforcing.

Those who tend to speak a little too softly should be aware of this and encourage themselves to speak so that people at the back of the room can hear them. If anything, err on the side of excess: better to speak at a volume that is slightly too high than slightly too low.

3. *Keep to a moderate pace—or slightly slower.* Speakers who are nervous often go too fast so as to get their presentation over with as quickly as possible. Speakers who are rather knowledgeable or who planned too much material can fall into the same trap for a different reason. Try to maintain an even, measured speaking pace. Think of it as a jog rather than a run. As with vocal projection, if you tend to speak too quickly when giving presentations, deliberately go slower than normal, even slightly slower than feels comfortable. Since an audience needs time to process your speech, it is better to go at a pace that is slightly too slow than one that is slightly too fast. Checking the time occasionally mid-speech can help inform your sense of pace under imposed or chosen time constraints.
4. *Adopt an open, relaxed stance in plain view.* Depending on the layout of the presentation area, some speakers, usually unconsciously, seem to want to hide—physically—from the audience: crouching behind a lectern, huddling against the back wall, and so on. All else being equal, you should make yourself visible in plain view of the audience. If you need to stand at a lectern to work a computer or consult your notes, try to stand in such a way that the audience can see a good deal of you naturally, even coming out from behind the podium to address the audience in a more engaging manner, in a TED Talk or town hall meeting style—which helps engage audiences.
5. *Consult notes only as needed.* Having detailed notes can make you feel comfortable that you will not get stuck in the middle of your presentation. This temptation is best avoided, however. For one thing, you could use your notes as a crutch, complicating the advice above about good eye contact. Additionally, seeking and finding specific content or cues you are looking for tends to be harder mid-speech. With the basic structure of your talk in mind, it will be relatively easy to deliver your speech extemporaneously and consult notes as needed when changing from one topic to the next or confirming, within a section, that you have covered everything you planned to.

Top Five Delivery Problems

Some problems with speech delivery are more common outside of kinesiology, such as, again, a failure to project sufficiently. Some problems, however, seem to be more prevalent in kinesiology than outside the discipline. Over the many years that I have taught communication to kinesiology students, certain flaws have predominated.

The behavioural quirks are exceedingly common, to be sure, and often speakers will be only marginally aware of them. But that does not make them any less distracting or make them seem any less awkward to an audience. If nothing else, such quirks, in an audience's mind, tend to take away from the credibility expected of a knowledgeable speaker, deflecting attention from the quality of your content. Great content poorly delivered yields *at best* a mediocre presentation. The following are, in my experience, the top five delivery problems in kinesiology presentations:

1. **Vocal fillers**. These are the *uhs* and *ums* that pepper many people's otherwise articulate speech, sometimes referred to as "vocal pauses" or "*dis*fluencies." Occasional vocal fillers are no big deal, but many people unconsciously include a distracting number while giving a speech. Vocal fillers are sometimes well-meant, reflecting a broadcaster's ideal of having no dead air, but you should avoid them. The antidote is to insert deliberate silent pauses when you are searching for just the right words. Far better to have a little silence than to fill it with noise.
2. ***Like* abuse**. This is the informal habit of not fully committing to the specific language you are using, often also for emphasis or exaggeration. In formal contexts, this suggests that your language use is sloppy or careless: "I like like her, but I don't like *like* her." Of course the word *like* is legitimate, but avoid using it to indicate merely approximating language. If an adjective you think of using is not quite right, choose a better word rather than simply inserting *like* before the imprecise one.
3. **Uptalk**. Uptalk is inflecting a statement as if it were a question. This is distracting and creates a bit of cognitive dissonance in the minds of the audience. Like vocal fillers, uptalk comes from a good place: an effort to be honest. The interrogative tone of uptalk often indicates your hesitancy to commit to the assertion you are making. But when you feel that you need to qualify your claim, do so with language rather than tone: not "This is the most important thing?" but "This is plausibly the most important thing," or whatnot.
4. *Mumbling*. Failing to speak distinctly and articulately is, again, a sign of informality that, in a formal context, seems careless or unskilled. In my experience, men are more prone than women to this tendency, but whether this is generally true and why this should be so is unclear. For those with a tendency to mumble, the remedy is to slightly overshoot precise-enough articulation. Better to over-articulate a little than to mumble.
5. *Inappropriate attire*. In this context, any outfit that avoidably distracts the audience from your presentation is inappropriate. Students sometimes dress up too much for presentations, even in-class presentations. But in my classes, the usual in-class error of many kinesiology students is

failing to dress up *enough*: presenting for credit in a baseball hat, T-shirt, sweatpants, and dirty sneakers, with little personal grooming besides, will make a poor initial impression on your audience. Aim for somewhere close to business causal: button-down shirts and blouses and plain sweaters or polo shirts rather than sweatshirts or T-shirts; slacks or medium to long skirts, or even dark, unripped jeans rather than shorts or sweatpants; and casual shoes or clean, dark sneakers rather than dirty, light ones.[6] For talks at conferences, more formal attire may be appropriate, depending on the subdiscipline and type of presentation. Asking more experienced students or professors for advice is always a good idea.

We can diagnose the general problem here: it is one of informality. Specifically, an *unintended* informality is frequently in effect in these cases. A student might not mean to mumble through their speech, abuse *like*, or alienate an audience by wearing sloppy clothing. That is just the way they normally speak and comfortably dress. Beyond an academic or professional milieu, such informalities are perfectly fine and even to be encouraged for social purposes. Here, excessive formality stands out as inappropriate. In formal contexts, however, it is crucial to be aware of and comport with whatever appropriate customs and expectations are in effect.

This is *not* to say that informality has *no* place in formal contexts, even when it comes to giving an academic presentation. For instance, if you are giving a talk on the role of playgrounds in keeping children physically active, it may be engaging to use in your title the Duke Ellington line "It don't mean a thing if it ain't got that swing" (Mills & Ellington, 1931). It might be paradoxical to suggest that there is a place for any informality in formal speeches, especially as many flawed presentations are marked by careless or unpolished conduct. The judicious use of informality in your presentation can, however, be a real asset. Just make sure it is a product of choice rather than carelessness, of design rather than disregard.

The Use and Abuse of Informality

In formal presentations, minimize unintended informalities, but consider selective informality to engage the audience.

VISUAL AIDS

We all know that visual aids can enhance a speech, providing text and visuals to reinforce and expand on what the speaker is saying. Along with this, however, is a virtue of visual aids that might be surprising: they can assist in the chunking and ordering of presentation material. That is, visual aids can be useful not just for *delivering* speeches but also for *creating* them. In a slideshow, for instance, individual slides lend themselves to information chunking. Slide design typically contains *one chunk per slide*, and the limitations of slide design can guide you in creating and ordering, in effective and focused ways, whatever text or images will accompany the presentation. If you find you cannot effectively fit all the material you have on a chunk on just a single slide, you should either dispense with the less important material or consider whether the too-big chunk should be divided into two separate slide chunks.

Pro Tip: Writing from Visuals

Use planned visual aids as a speechwriting tool.

Besides grouping information at one chunk per slide, keep in mind other complementary principles of design for slides or any other type of visual aids, including basic chalkboard or whiteboard work. Whenever possible, try to balance text and image elements, which will engage audience members who have different learning and attention styles. A typical slide might have, say, a heading, a handful of bullet points, and an appropriate image. Simplify these elements, including not only the number of images and words but also the font you choose, which is better the more it resembles block print rather than cursive writing.[7] Ensure readability and visibility by making the text and images as large as design limits allow without making the slides too crowded. You should also use high-contrast colours, such as black text against a white background or vice versa. In any series of points, try to use parallel language, phrasing items in the series using the same grammatical form (e.g., figure 11.1, in which each bullet point begins with a verb). In most cases, you should have a title slide as well as, for longer presentations, an overview slide that lays out the structure of your presentation. Make sure that you proofread slides and other

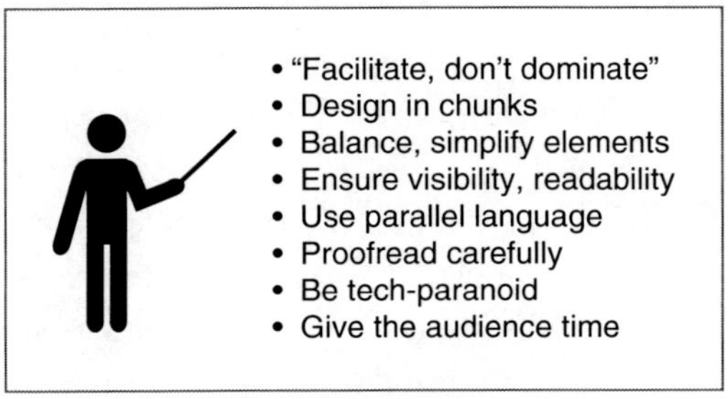

Figure 11.1. Visual Aid Tips
Source: Jason Holt

prepared visual aids for spelling, punctuation, and grammar mistakes, as the more of these there are, the less professional and credible you may seem.

View the in-presentation use of visual aids and their pre-presentation design strictly as *aids* to your speech rather than either a substitute for your speech or as doing most of the work. Visual aids are meant to help you present rather than to do the presenting for you. Avoid the temptation of using visual aids as a crutch. In slogan form, visual aids should "facilitate, not dominate," your presentation. Of a piece with this advice is, again, to be tech-paranoid, as I urge in other contexts as well. Whatever work you may have put into preparing, say, a slideshow, you should have contingency plans in place in case the technology fails. Be ready to use a chalkboard or whiteboard, or no visual aids at all, if your technology fails. If something beyond your control goes wrong, nothing is less professional than being thrown off and flustered, and nothing is more professional than proceeding smoothly. When things do go smoothly, however, make sure you give the audience more time to take in and think about your visual aids, because you, as the presenter, might go too quickly through material that is as unfamiliar to your audience as it is familiar to you.

IN-PERSON VERSUS ONLINE PRESENTATIONS

Presenting in person is all but essential in academic and professional kinesiology. Live in-person delivery allows the speaker to interact with the audience, including via verbal exchanges and picking up on nonverbal cues. Seeing a puzzled expression on several faces might prompt you to explain or rephrase your point. Presenting in class for credit might make you somewhat nervous, but the compensation is having a familiar venue and audience. You also have the advantage of a professor giving you clear or easily clarifiable expectations in the form of instructions. You can take cues as well from the instructor's own teaching style, along with what other students do well and not so well when they present. Avoid complacency, though. Do not assume, just because you are familiar with the setting and audience, that no extra preparation is required. Take the opportunity to test the technology beforehand, everything from computer hookups to light switches. Testing the technology and setting up early, if possible, is even more pressing if presenting in unfamiliar settings, such as at a conference. If the locale is new, find and explore the venue well in advance of the scheduled presentation time.

During the COVID-19 pandemic, the practice of remote online presentation became more common and important for speakers to master. Calling these presentations "virtual" is potentially misleading because you are *actually* presenting, just remotely. The virtuality lies in a simulation of the in-person presentation format.

Though, for obvious reasons, many aspects of preparation are similar to in-person presentations, there are a number of differences to manage, especially in the delivery of such presentations.

First, of course, you should become familiar with the basic features of whatever platform you are using, including how to access the platform itself and how to turn your webcam and mic on and off. If you are using visual aids, such as a slideshow, make sure to practise using them on the platform itself, however familiar you may otherwise be with the slideshow software. Certain functions that work well in a classroom setting, such as including links to video clips, tend to be glitchy when run through online platforms, so prepare for that. Tech paranoia here is also apt but has its limits, since some technical glitches—connectivity issues, computer crashes, and so on—might be simply beyond your control.

When it comes to the presentation itself, replace the actual balanced eye contact of an in-person presentation with *virtual* balanced eye contact. Look at the faces on the screen to assess nonverbal cues as appropriate, but remember that eye contact here is best simulated by looking not at the audience but rather at your webcam—in a sense, the opposite of the in-person experience. You should also be extra mindful of vocal distractors, especially vocal fillers, *like* abuse, and uptalk, along with potentially distracting or embarrassing features of the background. You should mute your mic before sipping water, clearing your throat, and so forth, especially if you are wearing a headset or have a mic otherwise close to your face. Unconsciously oscillating in a swivel chair is another potential distraction, so sit in a different kind of chair or be careful to sit reasonably still.

Online Presentation Tips

- Become familiar with the platform beforehand.
- Practise using visual aids on the platform itself.
- Maintain *virtual* balanced eye contact.
- Beware of vocal/background distractors.
- Mute your mic to sip water or clear your throat.
- Avoid or be mindful of swivel chairs.

This is all certainly a lot to think about when preparing and delivering a presentation, but with a measure of prudence and practice, this process can become more manageable, less burdensome, and perhaps even enjoyable.

BRIEF RECAP

In this chapter, we covered a wide range of topics pertaining to giving presentations:

- Public speaking
- Types of speech
- Engaging your audience
- Delivery tips and problems
- Visual aids
- In-person versus online presentations

CHAPTER 12

Speech Anxiety

CHAPTER OVERVIEW

In this chapter, we address the common problem of speech anxiety. Many people find themselves not just nervous about giving presentations but excessively nervous. We begin with a description of the phenomenon, followed by an analysis of its properties and the factors that influence it. We follow this with some practical theory about how to think about and manage such nervousness and its symptoms, and we end with a motivating analysis of how, in some ways, it is notably better to communicate in speech rather than writing.

GLOSSOPHOBIA

Comedian Jerry Seinfeld has a stand-up bit about people's fear of public speaking. He refers to a study that supposedly found people's number one fear to be speaking in public, with number two being death. "That means," so the punchline goes, "to most people, if you have to go to a funeral, you would rather be in the casket than doing the eulogy" (Seinfeld, 2020, p. 98).[1] All kidding aside, the fear of public speaking is a significant issue for many people. Seinfeld's joke is a little overblown, though it is somewhat plausible to think that a survey *could* produce those results. You have probably witnessed many examples of speechmakers who seemed too nervous to effectively present. It is a veritable cliché. Many people who appear composed in front of a crowd might well be nervous on the inside. When I walk into a class for the first time, despite my decades of experience teaching thousands of students, I myself will inevitably feel substantial trepidation. Even if you are a good performer or a seasoned veteran in front of crowds, the fear of public speaking will remain an issue to analyze and negotiate as you improve your communication skills.

The fancy term for this fear is **glossophobia**. Glossophobia, from the Greek-originating *glossa* for tongue—is the fear of *using* one's tongue. The term can be used broadly to signify a general reluctance to speak, in both interpersonal and small group interactions. But often it is meant more narrowly to indicate misgivings about public speaking specifically. If someone is comfortable speaking one-on-one and in small groups but becomes reluctant to speak in front of a crowd, they may be said to be glossophobic in the narrow rather than broad sense. Clearly, the narrow sense of the term concerns us most here.

We will attack glossophobia first by subjecting it to analysis to help us better appraise its characteristics and causes. We will then discuss some practically oriented theory to help give us a framework for managing both the feeling and the symptoms of speech-related nervousness. In addition to specific advice for symptom management in concert with refining one's presentation skills, I will also discuss various helpful mindsets—mantras, really—that variously help different people manage their glossophobia. Last, on the assumption that theory really can inform practice, we will delve into the philosophy of speechmaking and why, in certain ways, speech is superior to writing as a form of communication. Misgivings about giving presentations are often based, in part, on having a distorted sense of the format and its possible consequences.

ANALYZING THE PHENOMENON

An analysis of the phenomenon will help us get a better grip on what we are dealing with and so improve our chances of dealing with it successfully. In addition, the analysis will motivate a few practical measures for dealing with the psychology and symptoms of glossophobia.

Properties of Glossophobia

Part of dealing successfully with glossophobia is understanding some of its basic characteristics. These characteristics can be sorted into three groups: the neutral or objective side, what we may call the discouraging dark side, and the encouraging light side.

We begin with the neutral objective side. You should realize, first, that speech anxiety is, at the same time, normal and variable, each in two distinct senses. It may help to realize, if you have some measure of glossophobia, that you are not alone. Speech anxiety is entirely normal. It is normal both in being statistically common—a lot of people have it—and in that it is not so bad a thing to have, and so not something that reflects poorly on who you are or what you are capable of. Plenty of successful speakers have some measure of glossophobia, even while speaking, and so the fact that you feel it—however often, even to the level of distraction—does not imply that you are somehow doing a poor job or simply don't have what it takes. Speech anxiety varies across and even within individuals. It can have different causes and effects depending on the individual in question and their mindset at the time. Although many speakers are nervous about performing badly, others may be nervous because they believe the audience will be unreceptive to the message. Speech anxiety may manifest variously, with some people prone to blushing, shaking, vocal quavers, and so on.

Moving on to the discouraging side, speech anxiety tends to be significant and excessive. In terms of significance, it is a substantial issue that needs to be faced successfully by many speakers. For many people, it is not a mild and easily ignorable issue. If speech anxiety is an issue at all, it is usually a substantial one that speakers want to manage. It is significant also in the sense that the fear of public speaking signifies something: that the speaker is not well-prepared, perhaps, or that a lot is riding on how well the presentation is received. Speech nervousness is often not just substantial but, even in mild cases, excessive—that is, out of all proportion—to the implications of giving a mediocre or lousy speech. Glossophobics tend to overestimate, by a wide margin, the negative potential outcome of their speech, both beforehand and during delivery. In other words, to return to the Seinfeld joke, glossophobia is usually irrational, something that we feel we should be able to control more than we often do. Our goal should not be elimination but management. A rational approach suggests strategies to dial the anxiety down, not eliminate it outright.

As depressing as this darker side is, there is some reason to be optimistic, and so we turn to the lighter side. As I have already suggested, in most cases, even reasonably severe cases, the fear of public speaking is—and should be thought of as being—manageable by the speaker. The cases that require intervention by specialists in dysfunctional communication are rare. In any case, you should not be demoralized by your glossophobia. There are helpful ways of understanding the phenomenon and managing the symptoms, which is the goal of this chapter. Beyond manageability, you can even think of speech anxiety as a *good* thing, insofar as being nervous means that you care, and caring means you are motivated to do well, which itself means that, with the right tools, you can redirect that nervous energy more productively. Think of being nervous before a big presentation and then using that to fuel your preparation or motivate enthusiasm during the speech itself to help reinforce your message. Even doing isometric contractions while standing at the podium can diffuse a lot of nervous energy.

Speech Anxiety Is ...

- Normal
- Variable
- Significant
- Excessive
- Manageable
- Useful

MASS Factors

Judith Rolls (2007) identifies three broad potential sources of a speechmaker's nervousness: the material being presented, the audience presented to, and the speaker giving the presentation.[2] I will add another source of nervousness to yield a foursome that I will designate collectively by the acronym phrase "MASS factors," where MASS stands for material, audience, speaker, and stakes (i.e., what is riding on the speech). In many cases, these factors turn out to be double-edged, with different contexts resulting in different kinds of influence.

- *Material*
    - *Difficulty.* All else being equal, you tend to be more nervous presenting material that you find difficult to grasp. One reason for this is that you might suspect that the audience will be able to see through you and so judge you for your ignorance. On the other hand, presenting material that is particularly difficult for the audience—but that you know well, or even have some expertise in—could make you more relaxed. Either way, knowing your material well can help reduce speech anxiety.
    - *Interest.* Knowing that your audience is not really interested in what you have to say can increase your anxiety. You may want to do a good job, but if your content fails to engage the audience, it seems as if the deck is stacked against you. Conversely, if the audience *is* interested, you may become nervous simply because you believe they will scrutinize your speech more closely. Either way, it helps to be interested in the material yourself. Speakers who are not interested in the material they are presenting often lack the motivation to prepare properly and so, as they present, may find themselves more nervous than they should be. On the other hand, you might care so much about your message that you become over-anxious from the hope that your audience will receive it well.
- *Audience*
    - *Size.* As most people find it easier to speak to small groups than to speak publicly, the larger the audience for your speech, the more nervous you will tend to be. At some point, however, there can be a peculiar depersonalizing effect if a crowd becomes sufficiently large that it seems almost unreal, or at least not composed of individual people. Paradoxically, then, a sufficiently large crowd, for some people, might be seen almost as a single entity beyond their control and so not something to fear as much as they might a smaller but still sizable group.

  - *Familiarity.* Speakers tend to find that more familiar audiences are easier to present to. This complements a tendency to be more comfortable with acquaintances and friends than with strangers. Despite my many years of experience, I still get relatively nervous giving lectures to a class early in the term, although as the term progresses, my growing familiarity with the class puts me at ease in front of it. On the other hand, if you are giving a speech in a professional setting, your familiarity with members of a particularly demanding audience with intimidating levels of expertise may make you more nervous than you would be otherwise, not from fearing the unknown but from fearing the known.
- *Self*
  - *Personality.* It should be fairly obvious that different personality types will tend to experience different levels of speech anxiety as well as managing it with different levels of success. Introverts, for example, will naturally be more fearful in giving speeches than will extroverts. In fact, some extroverts are "natural" performers and find the prospect of presenting thrilling and something to relish, an opportunity to showcase their performative prowess.
  - *Mindset.* Irrespective of general personality type, the particular mindset you have when preparing and giving a speech can influence how anxious you are about the process. You may be reasonably self-confident and, in most situations, rise to the occasion with an appropriate level of self-trust and enthusiasm. But this may not be the case when dawn breaks on presentation day and you are standing at the podium. If you have recently received some bad news or suffered a personal setback, this can make it hard to tap into an often-available reservoir of self-confidence.
- *Stakes*
  - *Significance.* How much is riding on the speech will naturally affect your degree of nervousness in giving it. For instance, if you are a recent graduate embarking on a career as a professional kinesiologist, presenting to prospective colleagues is likely going be somewhat nerve-wracking, especially if you really want to get the job. You could be noticeably more nervous than you would normally be, because so much is riding on it. This applies even if you are generally comfortable in front of a crowd and are presenting material you know thoroughly. Remembering this can help reduce your anxiety when giving in-class presentations because, although such grading units obviously count for *something*, there is invariably less riding on them than your anxiety level would suggest.

- *Perception*. What matters, though, is not just what is riding on the speech but what you *think* or *feel* is riding on the speech, that is, its perceived significance.[3] If a lot is riding on the presentation, but you are unaware of this, it will not alter your anxiousness. It can't. You are usually aware of such facts, and so they, on their own, give you good reason to suppose you may be nervous. Yet even when little is *actually* riding on a speech, you may have the false impression that a lot *is* riding on it: that you might look foolish in front of classmates, that you will never live down doing poorly, or that your embarrassment will follow you throughout your degree program. This is a fear-based overestimate of the significance of speechmaking. What you fear could become an unforgettable disaster for your audience will almost surely be largely forgotten in relatively little time.

PRACTICAL THEORY

Turning from analysis to practical theory, we will explore ways of understanding speech anxiety that have practical implications for dealing with it. In other words, these approaches can reframe the speech situation in ways to reduce general anxiousness as well as pointing to helpful ways of dealing with specific and potentially distracting symptoms.

Task Orientation

Some theorists hypothesize that the source of speech anxiety can be a certain orientation toward the task of giving the presentation. This is sometimes described as the contrast between having a performance orientation and a communication orientation, though I will refer to it in terms of the distinction between *performative* and *informative* orientations. Someone with a **performative orientation** views the task of public speaking as a kind of performance, as actors or comedians do, with the purpose of entertaining an audience, where a lot depends on the audience's attitude. Someone with an **informative orientation**, by contrast, frames the task as a matter of conveying information. A performative orientation tends to be associated with stage fright, which lies at the extreme limit of speech anxiety, a debilitating psychological paralysis of the would-be performer. Because the informative orientation is more task-focused than speaker-focused, many theorists recommend it as an alternative to the unnecessarily nerve-wracking performative orientation.[4] The more you focus on the task itself, the reasoning goes, the less self-conscious you will be about giving the presentation.

No doubt there is wisdom in "taking yourself out of the equation" as a means of reducing speech anxiety, but several qualifications are in order. First, the distinction

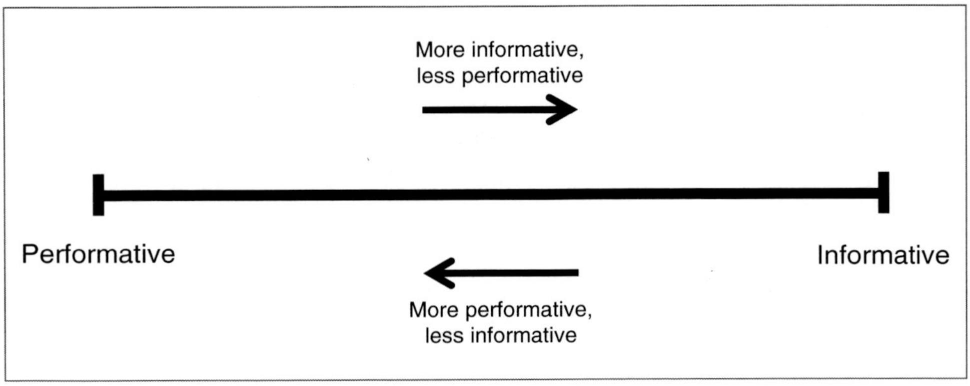

Figure 12.1. The Performative-Informative Spectrum
Source: Jason Holt

between performative and informative orientations should not be taken as a dichotomy, as if an individual either has one or the other exclusively. Rather, you should think of the performative/informative distinction as lying on a continuum between two extremes, with a good deal of middle ground where one may have complementary levels of performative and informative focus (figure 12.1). Think, then, not of adopting an informative orientation outright but of moving from left to right on this spectrum. Note, however, that you might find that your anxiety reduced by going toward the left rather than the right, especially if you have what I call **content anxiety**, which may arise from the sense that your content is not good enough or not sufficiently interesting. Alternatively, you may be more accustomed and enthused as a performer than as a lecturer, finding informative orientations somewhat disorienting. In some formal contexts, such as giving an official report to a committee, almost no degree of performance would be appropriate. Most formal presentations, however, including scholarly conference presentations, tend to go better with the incorporation of some performative elements, so long as they do not seem forced.

Reticence

In general, **reticence** is hesitancy or reluctance to speak, and, in a narrow sense, it has been applied to speech anxiety. You may be reluctant to say anything if you feel you have nothing to say, for instance. More central in this theoretical framework, however, is a reluctance that derives from a lack of know-how, a deficiency in public speaking skills themselves (Phillips, 1997). You might know, in great detail, *what* you want to convey or share in a speech but lack the wherewithal you need to be confident in *how* you present it. This is a plausible example of our natural tendency to shy away from tasks, even obligations, that we simply don't know how to do.

Apart from applying general advice about how to prepare and give presentations,[5] the most obvious way to acquire the necessary know-how is to *practise*. You should not just practise a particular speech beforehand but, more importantly, present often enough to become more accustomed to the situation. This includes being conversant in whatever visual aid technology you might use and doing practice runs that incorporate your prepared visuals. It also includes an obvious point for formal presentations that students and some professionals neglect surprisingly often: practising correct formal pronunciation, especially of frequently used terminology the audience should expect you to know how to pronounce.

Here are just a few examples of common informal pronunciations to avoid when giving a formal presentation. This is *not* an exhaustive list.[6] It is not that such informal pronunciations are *wrong*—are *mis*pronunciations—but that they can be distracting to an audience in a formal setting.

- *Ask*: *s* precedes *k* in sound, as in spelling
    - aks ☒
    - ask ☑
- *Athlete*: two syllables, never three
    - ath-*uh*-lete ☒
    - *ath*-lete ☑
- *Et cetera*: *t* precedes a soft *c* in four (or acceptably three) syllables
    - ek *se*-tra ☒
    - et *se*-tra ☑
    - et *set*-er-a ☑
- *Especially*: no *x* sound
    - *ex*-specially ☒
    - *es*-specially ☑
- *Library*: pronounce the first *r*
    - li-*bare*-y ☒
    - li-*brare*-y ☑
- *(Mis)pronunciation*: unlike verb, noun has no *ou*
    - (mis)pro-*noun*-ciation ☒
    - (mis)pro-*nun*-ciation ☑
- *Probably*: three syllables, not two
    - *prob*-ly ☒
    - *prob*-ub-ly ☑
- *Supposedly*: no *ab* in spelling or sound
    - suppose-*ub*-ly ☒
    - suppose-*ud*-ly ☑

Bear in mind, though, that some differences in pronunciation are conventionally acceptable. The word *skeletal* on its own or in *musculoskeletal*, for instance, may be pronounced with a stress on the first syllable and short second *e* or with a stress on the second syllable with long second *e*: "*skel*-e-tal" or "ske-*lee*-tal." Similarly, you should expect some differences in pronunciation given the speaker's region or country of origin. Where differences in pronunciation are unusual or suggest informality, it may be wise to soften them to minimize potential distraction. This also applies to regional quirks that have implications for spelling as well as pronunciation (appending *s* to the end of *somewhere* and *anywhere*, for instance).

A word of warning is appropriate, however. When we think of practice, we invariably think also of work ethic slogans such as "Practice makes perfect" and "Good, better, best." This is understandable. Some people might benefit from reminding themselves of the importance of a good work ethic. However, such phrases also prime us for a kind of unhelpful and even, in the public speaking context, unhealthy perfectionism. In fact, perfectionism can be counterproductive and can hamstring public speaking. You might notice that even famous orators, in recordings of their signature speeches, often exhibit notable quirks and imperfections. If you aim for perfection, you will tend to be thrown off mid-speech by noticing yourself making small errors. To counter such **speech perfectionism**, you should inure yourself to mistakes by embracing **speech "imperfectionism.**" If you *expect* little mistakes to arise during your presentation, when they do happen you will be far less likely to be thrown off your game. This is especially true for nervous symptoms that you are aware of often displaying. Expect them. So instead of "Good, better, best," I recommend a brand new adage: "Good, better, *good enough*." The complete variation, balancing a work and leisure ethic, would be "Good, better, good enough. Never let it get too rough, until your good is better, and your better good enough."[7] Practice makes *presentable*, not perfect.

Speech Imperfectionism

- Expect mistakes.
- Embrace "Good, better, good enough."

Helpful Mindsets

The perfectionistic mindset leaves speakers susceptible to being thrown by minor mistakes. This kind of negative thinking might be countered by what Phillips (1977) calls **rhetoritherapy**. The strategy here is to replace negative **self-talk** with positive self-talk (e.g., saying to yourself "You got this"). This is only one example of a helpful mindset to reduce the feeling and the symptoms of speech anxiety.

Along with positive self-talk, which many speakers find helpful, there are various other mindsets to help manage glossophobia. Not everyone will find all of these effective, but many mindsets will prove useful for some. It is a matter of experimentation. You might already use some such techniques in managing speech anxiety. Here are some mantras to keep in mind and try out to help you manage speech anxiety. Note that they are different from positive self-talk and might even seem starkly negative. They do tend to distract one from negative self-talk, however. They are also, interestingly enough, inspired by different philosophical theories, encouraging us to become more "philosophical" about speechmaking.

- **Realism** is the notion that you should focus on features of the speech situation that rationalize not being so nervous about it. For example, you may be nervous about giving a class presentation that is worth 5 percent of your grade. Focusing on how little the presentation actually counts as a matter of *fact* may help reduce your anxiety. You may feel vulnerable to looking like a fool, but so, too, is everybody else scheduled to present.
- **Stoicism** is a philosophy that encourages us to worry only about those things that we can actually control. This is captured to some extent in the familiar "Serenity Prayer."[8] Many people feel nervous when giving a presentation because they fear the audience will find them or what they have to say a waste of time. Although, to some extent, you might influence the audience's impression of you, it is wise to remember that what they think of you is beyond your control. This may ease your nervousness somewhat.
- **Existentialism** is the idea that *nothing really matters*, as many people feel when particularly depressed. The point of telling yourself that nothing really matters is *not* to depress you, although you might feel that way sometimes, but rather to put speech anxiety into perspective. Thinking to yourself that nothing really matters may make you feel that the comparatively small task of giving a presentation, even a major conference colloquium or job talk, is just not worth worrying about—not that the presentation *is* meaningless, again, but that it may help to *think* of it this way.
- **Solipsism** is a theory according to which the only things that exist are your mind and your current thoughts and experiences. The outside world, other people, and everything else that appears real are nothing but figments of your imagination. If you are worried about how the people in the audience will take your presentation, thinking that they don't really exist—though hardly true—can help make you less fearful of the encounter.

- **Buddhism** implies not that the audience doesn't exist, though this is also implied, but that *you yourself* don't really exist.[9] You might not be comforted, and might even be *discomforted*, by the thought that the people in the audience are unreal: a scary scenario. You could find it comforting, then, to think not that the audience doesn't exist but that *you don't* really exist! If you fear what the audience will think of you or what you say, thinking that you are not real may put you at ease, since there is no one there to suffer possible embarrassment.

The point, again, is not that any of these views are necessarily *true*, but rather that they might, as mantras, help reduce the substance and the symptoms of speech anxiety. We are often told to "be philosophical" about things that cause negative emotions, thus reducing their impact, and that is what my advice amounts to. Note, though, that we are splitting the difference: some thoughts that we find lessen our speech anxiety seem quite negative, such as the existentialist mantra. Remember that the efficacy of these mantras will vary depending on the context and the speaker alike. You should experiment with these and other mantras to find what works for you.

Symptom Management

When it comes to managing specific symptoms, the first step is to identify them, to be aware of what they are. You cannot successfully manage symptoms of which you are not cognizant. One strategy is to employ countermeasures, specific actions intended to replace or interfere with distracting symptoms. For instance, if you find that your leg shakes while you give a speech, you can try isometrically contracting the limb muscles for a span of time (say, 10 or 15 seconds) and then relaxing, contracting again if the symptoms return. Similarly, if you find making eye contact with anyone in the audience, even for a split second, extremely difficult, try focusing your gaze on the back wall, or look *almost* at people, at their shoulders or just over their heads, for instance, at empty seats between people, moving your gaze not person-to-person but seat-to-seat. If you find you have a vocal quaver, use that nervous energy to project *through* the quaver, since raising the volume of your voice may smooth out the tone.

Other strategies include varying the symptoms or interrupting them as they occur. If you tend, for instance, to oscillate when giving a speech, shifting side to side or front to back, which can be highly distracting, try mixing it up: shift forward and back, then side to side. That will be less distracting, and the more you vary such symptoms, the more it will seem almost natural and even an inclusive form of self-presentation to the audience. If you tend to clasp your hands or shove a hand into your pocket, acknowledge this tendency, but mix it up. Put a hand in your pocket, but then take it out after a few seconds and put your other hand in your other pocket. If you find you fold your arms in a closed posture, deliberately move to an open posture with arms held wide for a few second before allowing your posture, after a while, to close up again.

There are many other specific symptoms not covered here, and not all symptoms may be treatable. If you have a tendency to blush, for instance, there is little you can do about managing that symptom other than just feeling less nervous, which may be beyond your control while you speak. Whatever your nervous symptoms are, however, shifting focus away from the symptoms themselves—which can be *more* distracting to you than to the audience—is always a good idea. This coheres with the above advice to shift from a performative to an informative orientation in giving presentations. Shifting your focus away from yourself and your own imperfections and toward the task itself can, almost magically, reduce both your anxiety and its symptoms.

Symptom Management Strategies

1. Use countermeasures
2. Vary symptoms
3. Shift focus to task

SPEAKING VERSUS WRITING

Being philosophical about public speaking just means putting it in perspective. When we are too nervous about giving speeches, we tend to think too much about the consequences of the speech rather than focusing on the speech itself, the product rather than the process. If we appreciate the process more, we might better enable ourselves to manage speech anxiety.

In that spirit, I conclude with a view expressed in an ancient text about the relative merits of speaking to writing, of speech to text. The view is given in a work called the *Phaedrus*, written by the ancient Greek philosopher Plato (2005). Characteristic of Plato's work, the central figure of the *Phaedrus* is Socrates, Plato's real-life mentor. In conversation with the eponymous Phaedrus, Socrates expounds on speech as superior to writing as a form of communication. True to this philosophy, Socrates—a great philosopher—did not write his philosophy down, preferring to engage people in dialogue. The situational irony here is that Plato is one of the great writers in the Western tradition, and in a famous text two and a half millennia old is *writing* that writing is not the best way to communicate.

Both Socrates and Plato (2005) consider not only writing to be inferior to speech, but also *set* speeches, that is prewritten scripted speeches, to rank below the "living, animate speech" of true discourse (276a5). Plato sees discourse as superior to writing and set speeches alike in virtue of the following features:

- *Speech is better for memory.* Writing is often thought of as an aid to memory. That is why, for instance, you take notes in class. For many of us, however, writing can be a crutch. The more we rely on text, the less we exercise our memory muscles, and the more they atrophy. This problem has arguably become even more pronounced in the internet age.[10] Likewise, we tend to remember things *said* more than we do things *read.*
- *Speech is more interactive.* Since speech is a dynamic process compared to the fixed text of a piece of writing, we can interact with speakers in ways that we cannot with texts. A speaker can answer questions and provide further explanation, clarification, and examples, for example. A speaker can defend what they have said against people who have attacked it face to face. A speaker can also tailor a message to the members of the audience. By contrast, a text, on its own, can neither answer questions, explain further, nor defend itself better than it already does in its final form, much less alter itself according to the specific needs of different members of its readership.
- *Speech is natural.* Literacy, for all its virtues, is artificial. In most cases, we learn to speak and understand spoken language first, and it tends to be something many of us are more at ease with, although public speaking is a notable exception. One can be a master of language, even a genius, without being *literate*—without being able to read or write at all. Literacy is dependent on language itself in its more basic, spoken form. To that extent, then, it is both literally and in terms of people's comfort less natural.
- *Speech is ephemeral.* Writing, by contrast, is longer-lasting. That is why we can read Plato, who wrote 2,500 years ago, and know about Socrates from such writings. We cannot talk to Socrates, however much we may want to. So why would the fact that speeches are comparatively short-lived be a point in their favour? The answer, in a way, is straightforward. In most speech situations, you may communicate naturally and more openly because of a lower "**regret potential.**" A piece of published writing, for instance, will likely be around a lot longer to embarrass a writer who has made a mistake.[11] But in most cases, even when giving a presentation, one's errors are much quicker to fade—because they only exist in the minds of you and your audience.[12]

Making Speech "MINE"

Speech, relative to writing, is comparatively ...

- *M*emorable
- *I*nteractive
- *N*atural
- *E*phemeral

For Plato, then, the superiority of speech over writing has implications for what makes for better or worse speeches and better or worse writing. Speech is best when it is *least* like writing, and writing is best when it is *most* like speech. Good speech is interactive rather than closed off, dynamic rather than fixed. Bad speech is either like writing or constrained by writing in the form of a set speech. The best sort of writing, likewise, will be that which emulates speech, seeming at least more interactive and natural. It should be no surprise, then, that Plato wrote his philosophy in dialogue form. Were he alive today, Plato would disfavour transcriptions and video recordings of speeches. Likewise, and perhaps surprisingly, he might embrace *texting* as the best variety of writing, given its speech-like interactivity. Whatever irony we see here, Plato's view can help us to better appreciate speech and why even a formal presentation should be seen as a valuable, and otherwise unavailable, opportunity to communicate. May you now be better equipped to seize it.

BRIEF RECAP

In this chapter, we discussed the all-too-common problem of speech anxiety:

- Glossophobia
- Analyzing the phenomenon
- Practical theory
- Speaking versus writing

Notes

CHAPTER 1

1. Where one communicates with oneself at a different time, we could interpret this as involving two different people, assuming that who you are varies over time, that, for example, Holt at time 1 (writing a reminder) ≠ Holt at time 2 (reading the reminder). Such philosophizing about personal identity starts to become of more obvious practical relevance as the temporal gap widens.
2. I have put *olfactics* in scare quotes to indicate that Dobkin and Pace (2006) do not actually use this term. Rather I coin it here for consistency: *olfactics* is to *olfactory* as *proxemics* is to *proximity* and so forth.
3. Dobkin and Pace (2006) describe various relationships between verbal and non-verbal signals: accenting, complementing, contradicting, regulating, repeating, and substituting.
4. This problem is related to but distinct from what philosophers call the problem of other minds, where what is doubted is not the content of other minds but whether others *have* minds at all.

CHAPTER 2

1. The *Tao Te Ching* is an ancient work of Chinese philosophy written in verse and attributed to the poet Lao Tzu.
2. This seems obvious, but sometimes test stress, for instance, can lead students to forget simple things like bringing pencils or pens (and backups!) to a handwritten exam.
3. One advantage is that rehearsing different ways of phrasing thoughts keeps your ability to express yourself sharp and supple.
4. In the arts, analogously, it is sometimes said that you only discover what you mean to say by expressing it: in a poem, on a canvas, in music, or what have you. For details, see Collingwood (1938, pp. 109–110).
5. Occasionally I receive an email from a student with the salutation "Dear Professor Hot" rather than "Dear Professor Holt." This is a most unfortunate typo.
6. In the second case, the passive voice is used in the very assertion that it should never be used. The active/passive voice distinction is explained later.
7. The fourth edition is cited here. The original, by Strunk alone, was printed just over one hundred years ago.

8. In a poem, for instance, the unit of composition will be the stanza or individual line. For texts, the unit of composition may be the sentence or sentence fragment.
9. Many style guides, including Strunk and White, are clear that the passive voice is sometimes preferable, even necessary. However, people often take this advice as applying *absolutely* in all cases—as a strict rule rather than a rough guideline with clear exceptions.
10. Note that although the second option avoids using *we*, it also suggests that the author could be describing other people's activities, not their own. This impression is helpfully diminished by the passive voice. Since the trials themselves are the likely topic in question, they also serve as the appropriate subject of the sentence.
11. Think of the independent variable(s) as given or manipulated conditions and the dependent variable(s) as measured outcomes.
12. This outline follows the laboratory manual of the School of Kinesiology at Acadia University (2021). Your own required format may vary depending on your institution and department.
13. This seemingly trivial notion derives from a sophisticated account of tragedy originally given about two and a half millennia ago in Aristotle's *Poetics* (1999).
14. As media theorist Marshall McLuhan (McLuhan & McLuhan, 1988) might have put it, email not only *enhances* traditional letter writing (making delivery super-efficient) but also *obsolesces* it (by overtaking handwritten "snail mail").
15. Course syllabi usually tell you how to proceed in such cases.

CHAPTER 3

1. An exception to this is your institution's academic integrity policy, which should be available in your academic calendar and followed scrupulously. For more, see chapter 5 in this volume.
2. This formulation occurs on the inside front cover and so is not paginated.
3. Here I follow Freeland's (2001) helpful discussion of Baudrillard (1994).
4. A particularly striking and sad example is addiction to internet pornography.
5. An exception here is when gender differences are the subject of discussion, as with "Women's sports are often less popular than men's."
6. The derivation: *sportitude* = *sportsmanship* + *attitude* with connotations of *fortitude*.
7. As a matter of fundamental individual liberty, diverse sexual orientations arguably should be respected irrespective of whether they are subject to personal choice. Perhaps for some, if not most, people there is an element of choice involved—and why not?

8. Preferences here matter too. Last time we spoke, a Mi'kmaw acquaintance of mine was clear that *she herself* preferred to be called "Native" rather than "Indigenous" or "Aboriginal" despite *Native* no longer being the generally preferred term. If I were to refer to her as "Native" in deference to her preference, it would be wise, especially in formal contexts, for me to emphasize the subject's stated preference for such nonstandard use.
9. Indeed, sometimes your source *is* your subject, as when you do a critical analysis or review.
10. $E = mc^2$ is Einstein's famous formula stating that energy equals mass times the speed of light squared.
11. To be fair, the correspondence account of truth, though intuitive, is open to challenge. For our purposes, however, we will take it for granted.
12. Adapted from Williams and Colomb (2012, p. 124). Note that the gender-neutral pronoun *they* indicates either that we do not know Pat's gender or else that we know their pronouns are *they/them/their.*

CHAPTER 4

1. Fun fact: *Plato* is a nickname meaning "broad-shouldered," which the philosopher had earned as a wrestling champion.
2. Sometimes a single instance may serve as a representative sample, on the assumption that the case is *typical* (e.g., the structure of my brain is probably representative of the structure of most human brains).
3. This is my term for what I take as distinct from the moralistic fallacy: taking *ought* to imply *is* or *ought not* to imply *is not.* First, the moralistic fallacy seems limited to the optimist's version of the fallacy. Second, norms besides moral ones are also in play: rational, conventional, aesthetic, and so on, and so *normative* is ultimately a better counterpart to *natural.* Similarly, the naturalistic fallacy could involve inferring either *ought* or *ought not* from *is*, depending on one's general bias, if any, toward how reality unfolds.

CHAPTER 5

1. This advice applies no less to essay questions written on tests than it does to take-home essay assignments.
2. In the next section, I will directly address the qualitative assessment of potential sources.
3. The term *fake news* has recently been weaponized to discredit more traditional news sources for ostensibly political reasons. This is a remarkable shift in meaning given its original use in the entertainment industry for certain types of satirical political comedy (see e.g., Holt, 2007b).

4. As of this writing, for example, the Wikipedia entry for "Existentialism" (note 87) renders the subtitle of one of my own essays (Holt, 2007a) "Where *Do* the Paths of Glory Lead*?*" [emphasis added], not "Where the Paths of Glory Lead."
5. Be careful: finding an error does not automatically imply unreliability. If you have a keen eye, you might catch the occasional typo or other mistake even in a top peer-reviewed journal article or scholarly book. Reliability does not imply perfection. The more numerous and glaring the errors, however, the more likely the source is unreliable.
6. Treat the designation .biz similarly.
7. Adapted and condensed from Heckman (2007, pp. 24–26).
8. For details on how to document sources properly, see chapter 6 in this volume.
9. Students fresh out of high school are sometimes locked into the five-paragraph model, which implies *exactly* three paragraphs for the main body. In essays written at the university level, this usually results in unwieldy, unconscionably long paragraphs. I strongly recommend that, unless you are explicitly instructed otherwise, you abandon such rigid models. Use as many paragraphs in your main body as, given the assignment parameters, your considered outline requires.
10. More general advice on outlining is provided in chapter 2.
11. You have probably noticed that, in this scenario, since the essay has yet to be written, the talk will be extemporaneous, that is, prepared but unscripted. Chapter 11 suggests reasons why such a format is often more effective, even in formal settings where a script is available.

CHAPTER 6

1. Pro tip: writers often use hyphens (-) for page ranges in APA, which is usually not flagged as erroneous, even though in most contexts it is the en dash (–) that is strictly correct.
2. Such details are often available in documentation "quick guides" available on many university library websites (in this case Acadia University, n.d.). These tend to be more easily accessed and used than the style manuals available in your library's reference section.
3. Note that for short quotations, page references precede the punctuation, yet in block quotations they follow, indicating that they apply to the whole passage and not just the last sentence found on that page.
4. This error often pairs with no hanging indents (see below), as spaces visually separate entries the way hanging indents do.
5. I recommend that students avoid using automatic citation generators, which often omit data or provide them in the wrong form. At the very least, this crutch should be double-checked to prevent such mistakes. Sometimes citation generators yield little more than name, title, and URL.

6. This is the shortened form for the second and subsequent Chicago-style references to a given work. For a first reference, complete citation data are required except for the source's full page range (in lieu of which you should indicate the specific page or pages for the specific reference).
7. This is a noted form for CSE citations. There is also an in-text form similar to APA.

CHAPTER 7

1. For instance, on President Trump's inaugural portrait—soon withdrawn—was an inspirational quote that read, in part, "no challenge is to [*sic*] great," which should strike anyone, politics aside, as unprofessional. The Latin word *sic* means "yes" and indicates here that the mistake was that of the author, not the quoter. The brackets (i.e., [ ]) indicate an interpolation that is not found in the original passage but has been added by the quoter.
2. When you run farther down the road (preferred), you arguably also run further *down* the road.
3. The popular culture aficionado in me is convinced that this convention was popularized by TV shows in the *Star Trek* franchise (TOS, TNG, DS9, VOY, etc.).
4. *Online* is an outlier here, as it is becoming increasingly accepted as both an adjective and an adverb for all things pertaining to the internet.
5. For those who like formulas, here is a good rule of thumb: verb + preposition − [space] = noun (or adjective).

CHAPTER 8

1. I wonder how many sentence-ending prepositions in the manuscript for this book survived the copyediting process.
2. Indeed, it can help to think of language in general and formal language in particular as a kind of game. Seeing linguistic rules in terms of different "language games" is a view associated with the philosopher Ludwig Wittgenstein (1973).
3. Strunk and White (2000) advise writers to "avoid a succession of loose sentences" (p. 25), but this does not mean avoiding them altogether. Judicious use of a loose sentence or two may often prove stylistically preferable.
4. Northey and Seijts (2018), among others, prefer the locution that the subject "must *fit with* the predicate" [emphasis added] (p. 242).
5. For the mood variation, replace *will* with *would* in this example.
6. I borrow the term from Gilbert Ryle (2000), though his main concern is what he sees as errant views of reality deriving from implicit miscategorization. He thinks, for instance, that a category mistake underlies the view that there is something more to the mind than bodily behaviour.

7. Sometimes even great writers will fall into this trap. Ernest Hemingway's *The Sun Also Rises* (1954), my favourite novel, contains this evocative sentence with a dangling modifier: "Walking across the square to the hotel everything looked new and changed" (p. 192).
8. Think of including conditional verbs in the antecedent as *redundant* given that the *If* already indicates possibility, as typically does the consequent. Think also that, because they are conditional, such verbs *depend* on conditions that need to be specified in an antecedent and so do not belong there.
9. The phrase "prepositional gesture" is my proposed term for the phenomenon and is not conventional.

CHAPTER 9

1. In the colourful, informal language of indie rock band Vampire Weekend, "Who gives a f*** about an Oxford comma?" (Koenig & Vampire Weekend, 2007/2008, track 2).
2. Elizabeth Taylor was an actor and celebrity notorious for her many marriages, including *two* to fellow actor Richard Burton.
3. Sometimes single quotation marks are the base standard, as with many publishers based in the United Kingdom. In such contexts, simply reverse my advice in this chapter on using double and single quotation marks.
4. The source here is Holt (2016b, p. 6).
5. Scare quotes may also be used more broadly to indicate nonstandard but not necessarily loose usage. In formal writing, for instance, a technical neologism might be put in scare quotes to flag the novelty rather than imprecision of a term.
6. The singular of *bacteria* is *bacterium*.
7. This famous soliloquy is from Act III, Scene 1 (Shakespeare, 1958, p. 111).
8. Warning: some word processors autocorrect an apostrophe at the beginning of a word into an *open* single quote, where the apostrophe is properly identical to the *close* single quote. You can correct for this by typing the full word first, and then inserting the apostrophe before deleting the extra letter(s).
9. I do use semicolons now, but extremely rarely. You would find it challenging to find one used (rather than mentioned) in this book.
10. This interpolation comes from Holt (2019, p. 59), quoting Mumford (2012, p. 75). Not every style requires the use of brackets to indicate lowercase/uppercase discrepancies at the beginning of a quoted passage. Indeed, it is omitted in accordance with Routledge publishing house style in Holt (2019). Such scrupulous attention to detail, however, is usually wise.
11. This phenomenon is different from that of the **mondegreen**, wherein we persistently mishear song lyrics (e.g., the Jimi Hendrix lyric misheard "'Scuse me while I kiss *this guy*").

CHAPTER 10

1. Not all students are particularly conscientious even about content and so sometimes give the impression of simply going through the motions of doing just enough to get the credit they need for their degree. This may be a misimpression, but it is unfortunate either way.
2. Students sometimes like to use a subject's first name, in this case *Wayne*. I would discourage this trend, even if it expresses fondness for the subject. It can be taken as disrespectful, or at least too familiar and so presumptuous.
3. Keen-eyed students may have noticed that the form of both citations in the quoted passage is different from that which I have used throughout this book. This demonstrates the importance of being faithful to the source material even when you are required to use a different style yourself. My choice of APA documentation for this book reflects that this is the most commonly used style in various subdisciplines of kinesiology. Another thing you may have noticed is the use of single quotation marks rather than double, as in this book, which is another case of being faithful to the source.
4. Where the common sound is a vowel, this is called **assonance**. Where the common sound is a consonant, it is **consonance**. In the case of *poor sport*, we have both assonance and consonance. In some contexts, though, *bad sport* could be a better choice for competing reasons.
5. To illustrate, a former student and I recently published an article (Whitten & Holt, 2021). One reviewer praised the main title, "Thin for the Win," for immediately engaging the reader.
6. The technical term for newly coined words is **neologism**. See Holt (2013) on how neologisms can be an amusing and effective way to engage an audience.
7. This coined term includes itself in its own scope of reference.

CHAPTER 11

1. You can transcribe an unscripted speech, but the transcript generated *from* the speech does not make it scripted (i.e., set down *beforehand*).
2. This advice holds even in scholarly domains where conventions may suggest otherwise. Many researchers at conferences will literally read drafts of their papers to their audience. The rationale in certain fields of scholarship could be that precise language is of paramount importance, but in my experience, such gains are noticeably outweighed by the loss of interactivity.
3. Because of their addiction-level distraction potential, in class I will often refer to smartphones as "dumbphones." The term seems to have caught on, although not in this sense but in reference to basic cellphones without "smart" features like Wi-Fi and GPS.

4. Although presented in the context of business communication, this advice is largely applicable to presentations in kinesiology and, indeed, most other formal contexts.
5. Some kinesiology students prioritize comfort *too much*, at least for in-class presentations, and dress inappropriately as a result. This comes up again later in the chapter as one of the top delivery problems to avoid.
6. This may seem, but is not in fact, inconsistent with the advice to dress comfortably. Likewise, dressing appropriately may seem to be a distraction by making you *stand out* among your cohorts. But as the contrast is with an inappropriate, sometimes frustrating practice, the effect on an audience (especially professors or potential colleagues) is liable to be one of relief.
7. Fonts that are sans serif (e.g., Arial) are best for visual aids, as serif fonts (e.g., Times New Roman) are a little harder to read, and script fonts (e.g., Freestyle Script) are harder still.

CHAPTER 12

1. This bit appears in slightly different forms in videos of various performances by Seinfeld.
2. Rolls presents these in a different order—speaker, content, audience—and with somewhat different wording (2007, pp. 33, 41, 46). She also uses the term *cause* whereas I opt for *factor*, which is more open-ended.
3. I am using *perception* here to indicate a subjective sense of reality, a sense that may be false, rather than as a success term (i.e., in contrast to *misperception*).
4. See, for example, Motley and Molloy (1994), MacIntyre et al. (1997), and Rolls (2007).
5. As offered, for example, in chapter 11.
6. See also the "Tricky Singular/Plural Pairs" section in chapter 7.
7. Adapted from Holt (2021, p. 58).
8. Specifically the line "grant me the serenity to accept the things I cannot change." The prayer exists in various forms and was written, or at least popularized, by Reinhold Niebuhr, although its original authorship seems to have been a matter of some dispute (Crouter, 2010).
9. This is associated particularly with the Zen notion of *mushin* ("no-mind" or "no-self").
10. See, for instance, Carr (2020).
11. In my own case, the most embarrassing mistake I am cognizant of making in print—so far—is misquoting the famous last line from the movie *Chinatown*. Had I made this mistake during a lecture or conference presentation, the error, if noticed, would have started fading from memory—mine and other people's—fairly quickly.
12. This obviously fails to hold when speeches are transcribed or otherwise recorded. I return to this point later in the chapter, but note that transcriptions of a speech will omit most, if not all, of a speaker's anxiety symptoms.

Glossary

40+ Rule: quotations greater than or equal to 40 words should be set off as block quotations (APA style)

Abductive reasoning: inference to the best explanation

Abstract: summary of a longer work, often a one-paragraph essay summary

Acceptability: a condition for cogency in which the premises come from a sufficiently reliable source

Active voice: property of a sentence the subject of which is also the agent of the described action. *See also* passive voice.

Agents: the subjects involved in communication

Agreement: proper grammatical or compositional fit between different linguistic elements

Ambiguity: a quality of prompting of two or more different interpretations. *See also* vagueness.

Analogical support: similarity-based reasoning

Antecedent: the *if*-clause in a conditional (if-then) sentence. *See also* consequent.

Appeal to ignorance: the fallacy of citing a lack of evidence for a proposition as evidence for its negation or contrary

Appeal to popularity: the fallacy of inferring that a belief is true simply because it is common or that a thing has value simply because it is popular

Argument: set of statements, all but one of which are intended to support the other as rationally acceptable

Artifactics: object-based nonverbal cues

Assonance: sameness or similarity of vowel sound. *See also* consonance.

Attribution bias: the tendency, in explaining behaviour, to interpret ourselves too charitably (self-serving bias) and others uncharitably (other-disserving bias)

Aura: a special quality of unique artifacts or people undermined by the reproduction of their images

Author-date: type of in-text referencing requiring surnames and publication dates

Author-page: type of in-text referencing requiring surnames and page numbers

Block quotation: longer quotation set off by full indentation rather than quotation marks

Brackets: square parentheses used to indicate (1) content added to quotations or (2) meta-parentheses

Buddhism: in this context, a speech anxiety mantra according to which you do not really exist

Bullshit: statement made without any regard for the truth

Category mistake: faulty categorical implication produced by grammatical error

Channel: sensory medium through which messages are conveyed

Chronemics: time-based nonverbal cues

Cogency: property of good arguments wherein the premises are acceptable and the reasoning sufficiently strong

Coherence: property of a passage that is (1) intelligible or (2) unified, well integrated

Cohesion: the sense of flow created by linking adjacent sentences

Comma splice: improperly linking two independent clauses with a comma

Communication: the sharing of meaning via information transmission

Composition: creating messages in speech or writing

Conclusion: in an argument, the statement meant to be justified by the premises. *See also* premises.

Conditional sentence: sentence with an if-then structure

Conductive support: cumulative evidence–based reasoning

Consequent: the *then*-clause in a conditional (if-then) sentence. *See also* antecedent.

Consonance: sameness or similarity of consonant sound. *See also* assonance.

Content anxiety: fear that your presentation material is not good enough

Context: setting elements that affect meaning (e.g., culture)

Counterexample: an example that falsifies a universal generalization

Creative procrastination: doing work on one task as a means of avoiding another

Dangling modifier: modifies a logically improper explicit subject when the proper subject is left implicit

Deductive entailment: logic-based reasoning in which the premises, if true, guarantee that the conclusion is true

Demonstrative speech: speech meant to teach a procedure or skill

Dependent clause: clause that could not serve as a stand-alone sentence

Direct approach: stating conclusions at the beginning. *See also* indirect approach.

DOI: acronym for *digital object identifier*, a unique code for a work available in digital form

Em dash: a dash roughly the width of an *M*, used for emphatic asides, that can be typed as two hyphens without a space before or after

En dash: a dash roughly the width of an *N*, used for number ranges and to hyphenate compounds

Epigraph: a quotation at the beginning of a written work

Epiquotation: a misquote that persists in a culture because of its unique aesthetic appeal

Existentialism: in this context, a speech anxiety mantra according to which nothing really matters

Extemporaneous speech: prepared but unscripted speech

Fallacy: error in reasoning

Feedback: message in response to a message

Freewriting: writing without a preconceived plan, often to explore ideas or discover what one wants to say

Glossophobia: fear of public speaking

Grounds: a condition for cogency in which the premises provide sufficient justification for the conclusion

Haptics: touch-based nonverbal cues

Hasty generalization: the fallacy of inferring a generalization from a sample that is too small or otherwise nonrepresentative

Hyperreal: a property of simulations that have become more important than the reality they represent

Impromptu speech: spontaneous, unprepared speech

Inclusive language: terms that include members of marginalized or underprivileged groups and respect their self-referential preferences

Independent clause: clause that could serve as a stand-alone sentence

Indirect approach: saving conclusions until the end. *See also* direct approach.

Inductive support: statistics-based reasoning

Informative orientation: seeing speechmaking as communicating to the audience. *See also* performative orientation.

Informative speech: speech meant to convey information

Interpersonal communication: communication involving exactly two people

In-text references: specific citations, often enclosed in parentheses, given in the text rather than in notes. *See also* noted references.

Intrapersonal communication: communication with oneself

Justification: having sufficiently good reason for a belief

Kinesics: movement-based nonverbal cues

Knowledge: justified true belief

Lie: statement made to deceive someone about the truth

***Like* abuse:** informal practice of using *like* unnecessarily and noncommittally in speech

Loose sentence: a sentence with two or more independent clauses linked with a comma and a conjunction

Main body: the principal part of an essay, bookended between an introduction and a conclusion

Mass communication: communication involving large numbers of people

Message: what is conveyed or shared in communication

Meta-advice: advice about advice

Meta-parentheses: parentheses within parentheses, normally styled as brackets

Meta-quotation: quotation within a quotation, normally placed in single rather than double quotation marks (or vice versa)

Misplaced modifier: modifies a different linguistic item than the one intended

Mondegreen: persistently misheard song lyrics

Naturalistic fallacy: the fallacy of inferring value from fact

Neologism: (1) a newly coined word or phrase; (2) a new meaning attached to a standard word or phrase

Noise: channel-specific message interference

Nonverbal cues: either (1) unspoken or (2) nonlinguistic communication. *See also* verbal communication.

Normativistic fallacy: the fallacy of inferring fact from value

Noted references: specific citations noted as footnotes or endnotes rather than given in the text itself. *See also* in-text references.

Olfactics: smell-based nonverbal cues

Optimist's fallacy: the fallacy of inferring that something will happen simply because it ought to. *See also* pessimist's fallacy.

Other-disserving bias: the tendency in explaining behaviour to interpret others uncharitably. *See also* self-serving bias.

Paradigm: either (1) an ideal example or (2) a theoretical framework

Paralinguistic cues: nonverbal vocal modulation of spoken language (e.g., tone). *See also* vocalics.

Parallel structure/phrasing/language: property of a series where members have the same or similar grammatical form

Passive voice: property of a sentence the subject of which is not the agent of the described action. *See also* active voice.

Performative orientation: seeing speechmaking as performing for the audience. *See also* informative orientation.

Persuasive speech: speech meant to change beliefs or motivate behaviour

Pessimist's fallacy: the fallacy of inferring that something will not happen simply because it ought to. *See also* optimist's fallacy.

Plagiarism: presenting material as if it were one's own original work when, in fact, it is not. *See also* self-plagiarism.

Plausible: distinctly possible

Post hoc fallacy: the fallacy of inferring that one event caused another simply because the former preceded the latter

Predicate: the part of a sentence indicating what the sentence says about the subject. *See also* subject.

Premises: in an argument, the reasons offered to justify the conclusion. *See also* conclusion.

Prepositional gesture: use of weak or approximate rather than proper prepositions

Prewriting: writing such as note-taking and outlining in preparation for drafting

Principle of charity: in interpretation, giving authors and texts the benefit of the doubt

Productive distraction: break activities meant to inspire creativity and task-directed energy

Proposition: thought expressed by a statement

Proxemics: distance-based nonverbal cues

Public communication: communication involving more than seven people

Realism: in this context, focusing on how much less is riding on a presentation than one's speech anxiety would suggest

Regret potential: the degree to which an effort to communicate might embarrass you after the fact

Relevance: a condition for cogency in which the premises bear on the truth of the conclusion. *See also* topic relevance and truth relevance.

Reticence: hesitancy to speak

Rhetoritherapy: strategies for replacing negative self-talk with positive self-talk

Run-on fragment: sentence fragment long enough to be confused for a complete sentence

Run-on sentence: a sentence containing extra parts properly belonging in another sentence

Scare quotes: quotation marks used to indicate loose or imprecise usage

Self-plagiarism: submitting one's previous work for original credit, typically in a different course

Self-serving bias: the tendency in explaining behaviour to interpret ourselves too charitably. *See also* other-disserving bias.

Self-talk: talking to oneself aloud or silently. *See also* intrapersonal communication.

Sentence fragment: incomplete sentence

Set speech: speech that follows a script

Sharing paradigm: communication understood as the sharing of meaning

Small group communication: communication involving three to seven people

Social speech: speech meant to serve a social function

Solipsism: in this context, a speech anxiety mantra according to which the only things that exist are you and your current thoughts and experiences

Speech "imperfectionism": expecting and accepting that your speech will be imperfect

Speech perfectionism: thinking that any minor error will ruin a presentation

Spin: an attempt to influence the perceived significance of an accepted fact

Squinting modifier: ambiguous use of a modifier

Stoicism: in this context, a speech anxiety mantra according to which you should only be worried about things you can control

Subargument: an argument for a conclusion serving as a premise in a further argument

Subject: the part of a sentence indicating what the sentence is about. *See also* predicate.

Thesis statement: the unifying summary statement of an essay intended to be supported by the main body

Topic relevance: bearing on a particular subject. *See also* relevance.

Topic sentence: the unifying summary statement of a paragraph intended to be developed by the rest of the paragraph

Transmission paradigm: communication understood as the transmission of information

Truth: property of a statement that corresponds to reality

Truth relevance: bearing on the justification of a particular statement. *See also* relevance.

Uptalk: inflecting a statement as if it were a question

Vagueness: a lack of definite or precise interpretation. *See also* ambiguity.

Verbal communication: either (1) spoken or (2) linguistic communication. *See also* nonverbal cues.

Vocal fillers: inarticulate sounds made as pauses in otherwise articulate speech

Vocalics: voice-based nonverbal cues. *See also* paralinguistic cues.

Yoda syntax: saving the verb for the end of the sentence

References

Acadia University. (n.d.). *APA style 7th edition*. libguides.acadiau.ca/ld.php?content_id=35076308

American Psychological Association. (2020). *Publication manual of the American Psychological Association* (7th ed.). doi.org/10.1037/0000165-000

Aristotle. (1999). *Poetics* (F. Fergusson, Trans.). Hill and Wang.

Baker, S., & Gamache, L. B. (1998). *The Canadian practical stylist* (4th ed.). Addison-Wesley.

Baudrillard, J. (1994). *Simulacra and simulation* (S. F. Glaser, Trans.). University of Michigan Press. (Original work published 1981)

Benjamin, W. (2002). The work of art in the age of mechanical reproduction. In T. E. Wartenberg (Ed.), *The nature of art: An anthology* (pp. 173–181). Harcourt. (Original work published 1936)

Carey, J. W. (1992). *Communication as culture: Essays on media and society*. Routledge.

Carr, N. (2020). *The shallows: What the internet is doing to our brains* (Updated ed.). W. W. Norton.

Carroll, G. (Producer), & Rosenberg, S. (Director). (1967). *Cool hand Luke* [Motion picture]. Warner Bros.

Collingwood, R. G. (1938). *The principles of art*. Oxford University Press.

Crouter, R. (2010). *Reinhold Niebuhr on politics, religion, and Christian faith*. Oxford University Press.

Davis, D. (2002). The work of art in the age of digital reproduction (an evolving thesis: 1991–1995). In T. E. Wartenberg (Ed.), *The nature of art: An anthology* (pp. 317–328). Harcourt.

Dobkin, B. A., & Pace, R. C. (2006). *Communication in a changing world*. McGraw-Hill.

Dreyer, B. (2019). *Dreyer's English: An utterly correct guide to clarity and style*. Random House.

Existentialism. (2021, February 18). In *Wikipedia*. en.wikipedia.org/wiki/Existentialism

Frankfurt, H. G. (2005). *On bullshit*. Princeton University Press. (Original work published 1986)

Freeland, C. (2001). *Art theory: A very short introduction*. Oxford University Press.

Govier, T. (2004). *A practical study of argument* (6th ed.). Wadsworth.

Heckman, G. (2007). *Thomson Nelson guide to essay writing* (2nd ed.). Thomson Nelson.

Hemingway, E. (1954). *The sun also rises*. Charles Scribner's Sons. (Original work published 1926)

Holt, J. (2007a). Existential ethics: Where the paths of glory lead. In J. J. Abrams (Ed.), *The philosophy of Stanley Kubrick* (pp. 49–56). University Press of Kentucky.

Holt, J. (2007b). The Daily Show *and philosophy: Moments of Zen in the art of fake news*. Wiley-Blackwell.

Holt, J. (2013). Neologization à la Stewart and Colbert. In J. Holt (ed.), *The ultimate* Daily Show *and philosophy: More moments of Zen, more indecision theory* (pp. 298–308). Wiley-Blackwell.

Holt, J. (2014). Further inquiry. In J. Holt (Ed.), *Philosophy of sport: Core readings* (p. 297). Broadview.

Holt, J. (2016a). Epiquotation: Why we sometimes misquote stubbornly. *Evental Aesthetics, 5*(1), 4–14.

Holt, J. (2016b). Virtual domains for sports and games. *Sport, Ethics and Philosophy, 10*(1), 5–13. doi.org/10.1080/17511321.2016.1163729

Holt, J. (2017). Sport as art, dance as sport. *AUC Kinanthropologica, 53*(2), 138–145. doi.org/10.14712/23366052.2017.11

Holt, J. (2019). *Kinetic beauty: The philosophical aesthetics of sport.* Routledge.

Holt, J. (2021). Good better good enough. In *Poems for another time* (p. 58). Anaphora Literary Press.

Holt, J., Holt, L. E., & Pelham, T. W. (1996). Flexibility redefined. *Biomechanics in Sport, 13*, 170–175.

Howe, L. (2013). Gamesmanship. In J. Holt (Ed.), *Philosophy of sport: Core readings* (pp. 263–280). Broadview.

King, S. (2000). *On writing: A memoir of the craft.* Scribner.

Koenig, E. (Lyricist), & Vampire Weekend (Composers). (2007). Oxford comma [Recorded by Vampire Weekend]. On *Vampire weekend* [CD]. XL Recordings (2008).

Kuhn, T. (1970). *The structure of scientific revolutions.* University of Chicago Press.

Langan, J., Winstanley, S., & Jewell, C. (2009). *English skills with readings* (4th Canadian ed.). McGraw-Hill Ryerson.

Lao Tzu. (2008). *Tao te ching* (J. Star, Trans.). Tarcher.

Locke, J. (1836). *An essay concerning human understanding.* T. Tegg and Son. (Original work published 1690)

MacIntyre, P. D., Thivierge, K. A., & MacDonald, J. R. (1997). The effects of audience interest, responsiveness, and evaluation on public speaking anxiety and related variables. *Communication Research Reports, 14*, 157–168.

McLuhan, M., & McLuhan, E. (1988). *Laws of media: The new science.* University of Toronto Press.

Mills, I. (Lyricist), & Ellington, D. (Composer). (1931). It don't mean a thing (if it ain't got that swing). EMI Mills Music.

Motley, M. T., & Molloy, J. L. (1994). An efficacy test of a new therapy ("communication-orientation motivation") for public speaking anxiety. *Journal of Applied Communication Research, 22*(1), 48–58.

Mumford, S. (2012). *Watching sport: Aesthetics, ethics and emotion.* Routledge.

Northey, M., & McKibbin, J. (2007). *Making sense: A student's guide to research and writing* (5th ed.). Oxford University Press.

Northey, M., & Seijts, J. (2018). *Impact! A guide to business communication* (9th ed.). Pearson.

Office of Aboriginal Affairs. (n.d.). *The use of the terms Mi'kmaq and Mi'kmaw.* Government of Nova Scotia. novascotia.ca/abor/docs/links/Use-of-Words.pdf

Oxford University Press. (2021). Flexibility. In *Oxford English dictionary.* Oxford University Press. https://www-oed-com/Entry/71523

Phillips, G. M. (1977). Rhetoritherapy versus the medical model: Dealing with reticence. *Communication Education, 26*, 34–43.

Phillips, G. M. (1997). Reticence: A perspective on social withdrawal. In J. A. Daly, J. C. McCroskey, J. Ayres, T. Hopf, & D. M. Ayres (Eds.), *Avoiding communication: Shyness, reticence, and communication apprehension* (2nd ed., pp. 129–150). Hampton Press.

Plato. (2005). *Phaedrus* (C. Rowe, Trans.). Penguin.

Porhölä, M., Isotalus, P., & Ovaskainen, T. (1993). Communication-elicited arousal in public speaking, small group and dyadic contexts: Comparisons of change in heart rate. *Communication Research Reports*, *10*(1), 29–37. https://doi.org/10.1080/08824099309359915

Radford, G. P. (2005). *On the philosophy of communication*. Wadsworth.

Rolls, J. A. (2007). *Public speaking made easy* (2nd ed.). Thomson Nelson.

Russell, L., & Munter, M. (2014). *Guide to presentations* (4th ed.). Pearson.

Ryle, G. (2000). *The concept of mind*. Penguin. (Original work published 1949)

School of Kinesiology. (2021). *Kinesiology lab manual*. Acadia University.

Seinfeld, J. (2020). Public speaking. In *Is this anything?* (p. 98). Simon & Schuster.

Shakespeare, W. (1958). *Hamlet*. Dell Publishing. (Original work published 1603)

Shertzer, M. (1986). *The elements of grammar*. Collier Macmillan.

Starkey, S., & Zemeckis, R. (Producers), & Zemeckis, R. (Director). (1997). *Contact* [Motion picture]. Warner Bros.

Strunk, W., & White, E. B. (2000). *The elements of style* (4th ed.). Allyn and Bacon.

Whitten, L. J., & Holt, J. (2021). Thin for the win: Aesthetic bias and body image dissatisfaction in aesthetic sports. *Revue phénEPS-PHEnex Journal*, *11*(3), 1–10.

Williams, J. M., & Colomb, G. G. (2012). *Style: The basics of clarity and grace* (4th ed.). Longman.

Wittgenstein, L. (1973): *Philosophical investigations* (G. E. M. Anscombe, Trans.) (3rd ed.). Pearson. (Original work published 1953)

Index